EAS Publications Series, Volume 47, 2011

CNRS School in Astronomy for Professionals and Amateurs

Astronomical Spectrography
for Amateurs

Oléron, France
6–11 May, 2003

Edited by: *J.-P. Rozelot and C. Neiner*

EDP
SCIENCES

17 avenue du Hoggar, PA de Courtabœuf, B.P. 112, 91944 Les Ulis cedex A, France

Cover Figure

Design of the Lhires III spectrograph. The original version of this spectrograph was developed during the school. The Lhires III version is offered for sale by the Shelyak Instruments company created by amateur astronomers following the school.
© Shelyak Instruments

Indexed in: ADS, Current Contents Proceedings – Engineering & Physical Sciences, ISTP®/ISI Proceedings, ISTP/ISI CDROM Proceedings.

ISBN 978-2-7598-0629-4   EDP Sciences Les Ulis
ISSN 1633-4760
e-ISSN 1638-1963

© EAS, EDP Sciences 2011
Printed in UK

Preface

The CNRS astrophysics schools gather professional astronomers several times per year to discuss hot scientific topics. Since 2003 and every 3 years, at Oléron and then at La Rochelle, a CNRS astrophysics school welcomes, for the first time, amateur astronomers. That is how about 15 professionals and 35 amateurs, with already a good practice of astronomy, meet to exchange on topics and methods of observations. The goal is to have the passion for astronomy of amateurs converge with useful objectives for the scientific community.

This book is the result of lectures given during the first CNRS school gathering professional and amateur astronomers, held in Oléron (France) from May 6 to 11, 2003, and which was entitled[1]:

"Outils de l'astrophysique pour une coopération entre astronomes professionnels et amateurs."

During this meeting the contribution of professional to amateur work was mainly focused towards observing techniques, various possible common programmes or studies, scientific resources and methodology. The school was particularly focused on spectroscopy, a method to analyse light which allows a better knowledge of the physics of astronomical bodies. A cornerstone of professional astrophysics, spectroscopy now tends to develop among amateurs.

The goal of professional-amateur schools is to guide amateurs by proposing on one hand an update on astrophysics and on the knowledge of professional scientific needs, and on the other hand to define together a methodology for observing programmes. Moreover, professionals can realize during these schools what amateurs can bring to their studies and propose federative scientific programmes.[1]

Indeed, in spite of the means put together by professionals, with an increasing complexity, the sky remains vast. A number of observations can be executed by amateurs with small instruments and become part of larger professional campaigns. Compared to powerful professional telescopes, the instruments of amateurs are of course modest, but they have recording and measuring equipments with the same technology as professionals. In addition, size is not always a handicap if the observations are frequent and regularly scheduled. Finally, amateurs have a much bigger reactivity to follow a sudden and unexpected astronomical event than professionals who have to follow a long procedure before pointing a telescope to a chosen object.

Nowadays a large number of amateur astronomers are equipped with an instrumentation allowing to produce measurements likely to interest the community of professional astronomers. In particular, the use of CCD cameras has widely spread

---

[1]See http://www.astrosurf.com/aude/oleron

among amateurs and opened the road to already fruitful exchanges in domains as various as the discovery of supernovae, photometric follow-up of variable stars, meteorology of planets, occultations of stars by asteroids, mutual phenomena of Jupiter's satellites, etc. It is worth pointing out, in addition, that nowadays amateurs can access numerous 600 mm-class telescopes or telescopes that are smaller but highly automatised, which are very efficient for sky survey programmes.

In this context, the observation by amateurs takes a new dimension: from a simple hobby it becomes useful and contributes to a better knowledge of astronomy. Moreover, with this school, the astrophysics scientific community shows its willingness to open up and share science with a broad audience. The successive editions of the schools, in 2003, 2006 and 2009, have all been strikingly successful, and a forthcoming school is planned in 2012.

The successive chapters of this book will provide, to all those who have a true passion for astronomy, an idea of the state of the art in the various domains it tackles: the first chapter exposes the basics of physics of light. The following two chapters present respectively the spectrographs used in amateur astronomy and the treatment of the data obtained with these instruments. The following chapters (4 to 7) show examples of spectroscopic applications to the Sun, Be stars, planetary nebulae, and comets. This book will thus be useful for all, from the enlightened amateur to the professional.

DOI: 10.1051/eas/1147000

Contents

Astronomical Spectrography for Amateurs
J.-P. Rozelot and C. Neiner (eds)
EAS Publications Series, **47** *(2011) 1–37*

THE PHYSICS OF LIGHT FOR OPTICAL SPECTROSCOPY

A. Klotz[1]

Abstract. This chapter focuses on the fundamental origin of lines observed in astrophysical spectra. Some properties of the atomic world are qualitatively explained as an introduction to spectroscopy. General laws and rules are exposed and applied to celestial bodies.

Please note: it is possible to read this chapter quickly concentrating only on the sentences in bold type.

1 Photons and colours

The photon is an elementary unit for light. Up to now, experimental physics tends to prove that photons have no mass. As a consequence, a photon is not a material particle. In the vacuum, it moves at a velocity about of $300\,000\,\mathrm{km\,s^{-1}}$. This speed is usually symbolized by the letter c.

An electromagnetic wave is associated with each photon. Such a wave is constituted by two fields perpendicular to each other and to the trajectory of the photon: the electric field $\vec{E}$ and the magnetic field $\vec{B}$. Generally, from a photon to another, fields have any angular direction but some media can favor a direction. It is the polarization of the light, not discussed in that paper. A useful application of polarization can be read in the chapter "Spectroscopy of Be stars" by C. Neiner.

Light can be described as a wave and as a particle associated to photons.

The two concepts, wave and corpuscle, co-exist together, even if it is difficult to admit for our mind!

The electromagnetic wave, associated to a photon, has a characteristic called wavelength, symbolized by λ (lambda in the Greek alphabet). If we could take a photography of the wave, we would see a sinusoidal oscillation (Fig. 1). The distance between two consecutive maxima of fields is the wavelength. Photons,

---

[1] Institut de Recherche en Astrophysique et Planétologie, Observatoire Midi-Pyrénées (CNRS, Université Paul Sabatier), BP. 4346, 31028 Toulouse Cedex 04, France

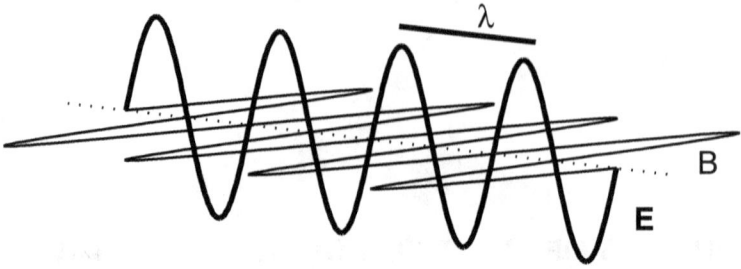

Fig. 1. Schematic view of an electro-magnetic wave. λ is the wavelength associated to the photon. The electric $\vec{E}$ and magnetic $\vec{B}$ fields are perpendicular to the path of the photon (symbolized by the dashed line).

visible to the human eyes, have wavelengths in the range 3000 to 7000 Ångströms (labeled Å hereafter).

One Ånsgtröm is equal to 10^{-10} meter. 3000 Å corresponds to the color violet and 7000 Å to deep red.

The Ånsgtröm is a unit that has been used for a long time by spectroscopists. In the range of visible light, tables and spectra are usually indexed with the Ånsgtröm unit although the nanometer unit (10^{-9} m) is recommended by the International System Unit committee. In this paper, we use the Ånsgtröm unit.

Instead of describing the oscillation by the wavelength, one can also use the frequency, symbolized by ν (nu in the Greek alphabet). If we could stand on a point of the photon trajectory, we could count the number of maxima of the field for a given duration. It is the frequency, expressed in Hertz (noted Hz). However, for the visible light, the wavelength is commonly used.

As the photon moves away, it has a kinetic energy. The photon energy is proportional to its frequency. In optical spectroscopy, the energy is usually expressed in electron-Volts (eV). 1 eV is about $1.6 \cdot 10^{-19}$ Joule. Another unit, cm^{-1} is also used in accurate laboratory spectroscopy tables. Keep in mind that 1 eV is about $8065.5\,cm^{-1}$.

Photons visible for human eyes have energies in the range 2–4 eV. The less energy the photons have, the redder they are.

Generally, spectrographs are used in the ambient air. In these conditions, measured wavelengths are slightly different from those measured in the vacuum. Wavelengths in tables established in laboratories are often quoted for so-called standard air (dry air, pressure 760 mmHg, temperature 15 °C and a volumic concentration of carbon dioxyde of 0.03%). In practice, for visible light, air-measured wavelengths are about 1.5 Å higher than if they were measured in vacuum.

2 The hydrogen atom

Atoms are elementary particles of matter in the objects of the known Universe. The hydrogen is the simplest atom. It is constituted by a proton (the nucleus) around which an electron can be found.

The mean distance between two atoms depends on the density of the material medium. In an atomic gas, at the atmospheric pressure, at temperature of $300\,\mathrm{K}$ ($20\,^{\circ}\mathrm{C}$), the distance is about $15\,\text{Å}$. In a molecule, the mean distance between adjacent atoms is about $1.5\,\text{Å}$. That last distance is considered as characteristic of a dimension of atoms (about $1\,\text{Å}$). The next section will show that this view is very (too) naive.

2.1 An atom seen closely

To analyze an atom, we must reduce the analytical tools (rulers, compasses, flashlights, etc.) to the dimension of an atom. But this is impossible because the tools will be atoms themselves and not instruments for measurements! This paradox can be generalized and any probe will perturb the atom that one wants to analyze. As a consequence, it is not easy to give a realistic view of atoms. We must exclude a sensory approach such as touching. One could imagine looking at the atom placed far away, but we must illuminate it and light can also perturb the atom.

Let us consider an hydrogen atom in its ground state, *i.e.* not excited. Niels Bohr, at the beginning of the XXth century, proposed that the electron is a point particle moving around the proton, as planets are in orbits around the Sun. If we could be close enough to the atom, reality would be far from this theory. As a sensory approach is impossible and as a probe cannot analyze it without any perturbation, we will use an imaginary instrument. This instrument should be a small detector of the presence of the electron (in a volume of about $0.001\,\text{Å}^3$ for instance) which does not perturb the atom. Driving this virtual detector around the atom, we are now able to map the locations where the electron is detected, as predicted by modern theories.

If the planetary orbit theory of Bohr was right, our tiny detector should detect the presence of the electron only in the trajectory of the orbit (a circle according to Bohr). It is not the case. Even at few Ångströms from the nucleus, the detector can measure a weak signal of the electron's presence. When we come closer to the nucleus, the electron's presence becomes higher. It implies that the electron is everywhere in space but it is more present near than far from the nucleus. After driving the detector around the the atom and scanning it at many distances, one can draw an image of measurements. The more the electron is present, the darker the color will be.

The Figure 2 shows the probability of finding an electron (the probability density), in the ground state, around a proton (it is an hydrogen atom). An analogy can be drawn for the probability density of the electron and a blanket of fog. In a blanket of fog, one can measure density variations from point to point

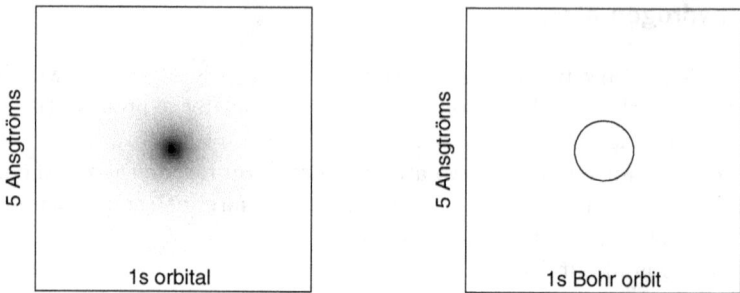

Fig. 2. Left panel is a view of the density probability to find an electron placed in the ground level of the hydrogen atom (1s orbital). On the right panel, the circle is the orbit described by the Bohr theory which is far from the reality.

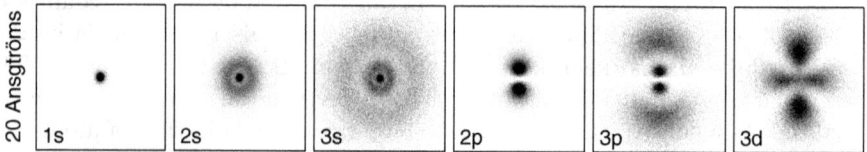

Fig. 3. Viewgraphs of the spatial probability distribution of an electron placed in the six lowest-energy states of the hydrogen atom.

but the fog is everywhere. Moreover, it is difficult to determine the limits of a blanket of fog. One can choose a low limit of density to determine a contour of the blanket but the fog can be found beyond this arbitrary limit.

In an atom, the electron cannot be localized in a point, nor be delimited by a finite volume.

Another analogy with a blanket of fog concerns the motion of the electron. In the Bohr theory, the electron is a point that follows a circular trajectory centered on the nucleus. In reality, the cloud of the probability density of the electron does not change with time. This does not mean that the electron does not move. A blanket of fog can be animated by internal motions without changing its global shape. In this case, it is a steady state, which is different from a fixed state. As a consequence:

A trajectory cannot be established for an electron at a steady state in an atom.

Inputs of external energies can put the hydrogen atom in other steady states. They are the excited states. The electron cloud extension of the density of probability can become very different. However, it cannot take any arbitrary form. An orbital is the mathematical function which defines the probability density in three dimensional space (Fig. 3).

When an electron is in an orbital, it has a very precise value of its mechanical energy. The mechanical energy is the sum of the kinetic energy, due to the motion of the particles, and the potential energy, due to forces describing the physics of the system (the electrostatic potential in the case of the hydrogen atom).

Quantification illustrates the fact that the mechanical energy of an electron cannot reach any arbitrary value when it lies in an atom.

3 Electronic states

An electronic state corresponds to the meaner in which electrons are placed in orbitals. A mechanical energy is associated to each state.

The mechanical energy of the ground state for an electron in an hydrogen atom has a value of about -13.6 eV. This state is linked to the principal quantum number $n = 1$.

The energy is negative when electron is bound to the nucleus. This remark is valid for every mechanical system, even if it is not governed by quantum mechanics: a negative mechanical energy always mean the system is bound. For instance, the Earth interacts gravitationally with the sun. Its mechanical energy is negative.

For a given energy, there can be many electronic states which have different probability densities. For each of these states, a second integer number, called the azimutal quantic number (noted l) characterizes the state in greater details. $l < n$ and n begins from one and l begins from zero.

A hydrogen atom in a given electronic state is defined by the pair of numbers (n, l). For instance, $(1, 0)$ is the ground state. To avoid confusion between the two numbers, it is preferable to use a symbolic notation where l is replaced by a letter: s for $l = 0$, p for $l = 1$, d for $l = 2$, f for $l = 3$ and alphabetic order for the next (g, h, etc.).

The electronic ground state for the hydrogen atom is noted (1s).

The Grotrian diagram is a graph where energy is reported *versus* families of quantum numbers (same l values in the case of hydrogen).

The four symbols s, p, d, f come from the XIXth century when experimental workers interpreted spectra without modern theories. They classified the observed lines into series according to subjective criteria: s means sharp, p means principal, d means diffuse and f means fundamental. At the XXth century these symbols were attributed not to the series but to the electronic states.

The energy levels become closer to each other as the principal quantum number increases. Levels tend to form a continuum from the zero value for the positive mechanical energy. In the ground state, the electron is strongly bound to the proton ($E = -13.6$ Ev). The atom is excited when a positive energy is added to the system. As a consequence, the electron climbs from level to level until it reaches the level of zero mechanical energy. At this point, the electron is no longer bound to the nucleus. It means that the kinetic energy of the electron

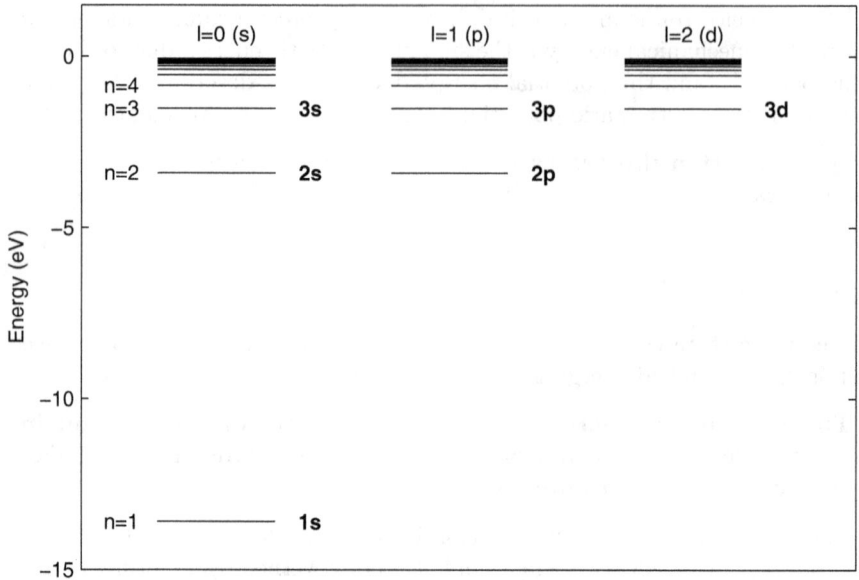

Fig. 4. Grotrian diagram for the hydrogen atom. Each horizontal line is placed at the corresponding energy of one state of the proton-electron system. The ground state is 1s. Only states for azimutal quantic number $l < 3$ are drawn. As $l < n$, states (1p), (2d), etc. do not exist.

(which always has a positive value) is higher than the potential attraction from the proton (which always has a negative value). Then, the electron becomes free and can go far away from the nucleus which becomes an ion. A free electron can attain any positive energy, hence the continuum (see Fig. 5).

The zero value of the mechanical energy is usually called the ionization limit. The ionization energy is the value of energy from the ground state to the ionization limit.

The ionization energy of the hydrogen atom is 13.6 eV.

4   Quantum numbers

The previous section shows the existence of two quantum numbers. These numbers are not empirical parameters. They come from basic Physics. One can find an analogy when comparing the energy diagram of the electron-proton and that of comet-Sun (Fig. 5).

For the electron, the bound states have any integer value of the quantum number n. For the comet, bound states are characterized by eccentricities lower than one (orbits are ellipses in that case). It is important to compare the role of n and the eccentricity. The eccentricity value depends on the initial conditions when

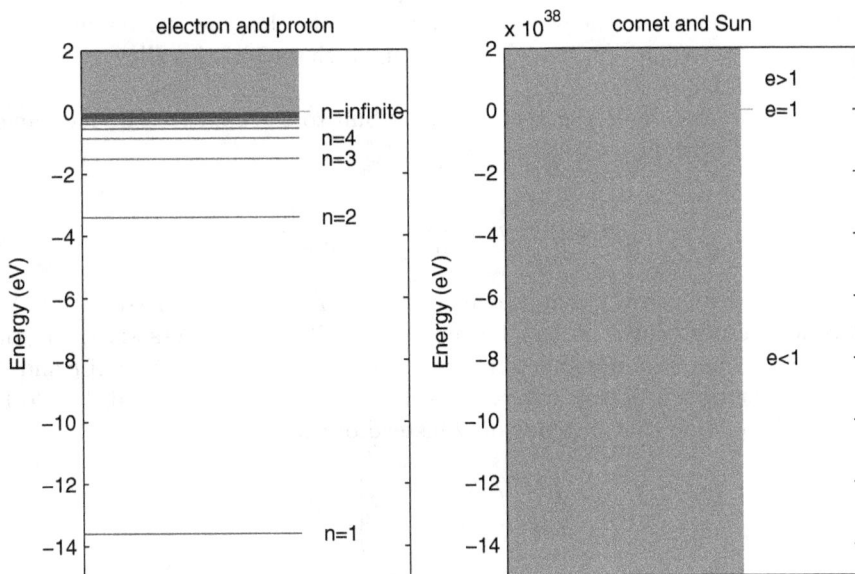

Fig. 5. *Left*: diagram of allowed energies for the proton-electron system of hydrogen atom. Energies are quantified for negative values. *Right*: for a gravitational comet-Sun system. All energies are allowed.

the Comet was placed in its orbit. For instance, we may explain the low eccentricity of the Earth's orbit by suggesting that it was formed gently and governed by the rotation of the protosolar nebula. Another example, concerns comets of the Oort cloud. These comets come from the more distant regions of the solar system where the potential energy is very low. These comets were formed at the peripherical regions of the protosolar nebula and are supposed not to be dragged away by its rotation motion. It implies that the kinetic energies of these comets are very low. The Oort cloud comets have very simple initial conditions because kinetic and potential energies are close to zero. This is the origin of their parabolic orbit because it can be demonstrated that the eccentricity value is rigorously one (parabolic orbits) when $E_{mec} = 0$.

The origin of the quantum numbers comes from the initial conditions allowed for the electron-proton system.

The previous approach is qualitative. This result can be demonstrate more rigorously.

We recall that the mechanical energy is the sum of the kinetic and the potential energies of the body considered in the system. In a hydrogen atom, the kinetic energy mainly comes from the motion of the electron and the potential energy comes from the electrostatic interaction between the electron and the proton. In the case of a comet and the Sun, the kinetic energy mainly comes from the motion of the comet and the potential comes from the the gravitational interaction between the comet and the Sun. These common characteristics allow us to begin the demonstration.

We assume that $\vec{r}$ is the vector linking Sun to the comet. The mechanical energy can be written:

$$E_{mec} = \frac{1}{2}m\left(\frac{d\vec{r}}{dt}\right)^2 - \frac{GMm}{\|\vec{r}\|} \tag{1}$$

where G is the universal gravitational constant, M and m are masses of the Sun and comet respectively. $\vec{r}$ can be replaced by its components (r, θ) in polar coordinates, it becomes possible to rewrite the equation as a function of r and the angular momentum L. Let $u = 1/r$. Substituting in Equation (1) leads to another differential equation that depends on θ instead of time:

$$E_{mec} = \frac{L^2}{2m}\left[\left(\frac{du}{d\theta}\right)^2 + u^2\right] - GMmu. \tag{2}$$

As the mechanical energy is constant in an isolated system, then $\mathrm{d}E_{mec}/\mathrm{d}\theta = 0$ and the derivative of Equation (2) can be put into the form:

$$\frac{d^2u}{(d\theta)^2} + u = \frac{1}{p} \tag{3}$$

where p is a constant term. This is a second order differential equation where $1/p$ is a particular solution and $u(\theta) = A\,cos(\theta - \theta_o)$ is the general solution of the equation without the second term. Replacing u by $1/r$ leads to the expression of polar motion:

$$r(\theta) = \frac{p}{1 + Ap\,cos(\theta - \theta_o)}. \tag{4}$$

This is the general equation of a conic where the product Ap is the eccentricity (noted e) and where the quantity $q = p/(1 + e)$ is the perihelion distance. As a consequence, (e, q, θ_o) are the initial conditions of the system. One can deduce the expression of the mechanical energy replacing expression (4) in Equation (2):

$$E_{mec} = \frac{GMm(e - 1)}{2q}. \tag{5}$$

In the case of the electron-proton system of the hydrogen atom, we cannot use Equation (1), because the trajectory is not defined (Sect. 2.1). However, it is possible to use the probability density ρ. The density is spread out over space (Fig. 3). At any point of space, located at the end of the $\vec{r}$ vector, the goal is to compute the $\rho(\vec{r})$ function which defines the probability of finding the electron at that place. Unfortunately, the calculation is not direct and it is necessary to use the wave function as an intermediate.

Physicists demonstrated, at the beginning of the XXth century, that one can define a complex mathematical function Ψ to verify $\rho = \Psi^*\Psi$ (where the Ψ^* denote the conjugated complex function of Ψ). Ψ is the wave function associated to the particle. Our physiological senses cannot detect a wave function and this is one of the reasons that quantum mechanics seems to be so abstract. From the laws of quantum mechanics, the mechanical energy has the form:

$$E_{mec} = \frac{\hbar^2}{2m\Psi(\overrightarrow{\mathbf{r}})} \left(\frac{d\Psi(\overrightarrow{\mathbf{r}})}{d\overrightarrow{\mathbf{r}}} \right)^2 - \frac{e^2}{4\pi\epsilon_o||\overrightarrow{\mathbf{r}}||} \tag{6}$$

where the first term is the kinetic energy and the second is the electrostatic potential. The symbol $\hbar$ means $h/2\pi$ where h is the Planck's constant for whose value is about $6.6 \cdot 10^{-34}$ J $\cdot$ s; m and e are the mass and the electric charge of the electron respectively. The kinetic energy term has a disconcerting form which strictly speaking has not be demonstrated. It is a postulate that comes from the equivalence principle which says that the behavior of the kinetic term must tend to the classical formula as the systems became larger.

Equation (6) is very different from Equation (1). As a consequence, the mathematical solutions are not the same. The solutions of Equation (6) use polar coordinates (r, θ, ϕ). It is possible to split them into two differential equations, one depending only on r and the other depending only of θ and ϕ. The mathematical form of the solutions can be written as:

$$\Psi(r, \theta, \phi) = R(r) \cdot Y(\theta, \phi) \tag{7}$$

where $R(r)$ are solutions of the differential equation depending only on the distance r and $Y(\theta, \phi)$ are solutions of the differential equation depending on the spatial variables θ and ϕ. Moreover, these differential equations are second-order which implies that solutions R and Y depend on the initial conditions. The numbers (n, l, m) are the set of initial conditions (m is defined as the magnetic number). They play a similar role to (e, q, θ_o) for the comet-Sun system previously studied. However, it is important to emphasize that (e, q, θ_o) can reach any real positive values while values of (n, l, m) can only take integer values, with the following limitations: $l < n$ and $-l \leq m \leq +l$.

We said (Sect. 2.1) that the electronic density (hence the wave function Φ) is in a steady state. It can be demonstrated that values of (n, l, m) must be integers to prevent the steady state waves from interfering with themselves. These integer values are at the origin of the adjective *quantum* for the mechanics applied in atomic scale physics. Finally, introduction of the solutions (7) in Equation (6) leads to the expression of the mechanical energy:

$$E_{mec} = \frac{me^4}{(4\pi\epsilon_o)2\hbar^2 n^2}. \tag{8}$$

The mechanical energy of the electronic states of the hydrogen atom depends only on the principal quantum number n. As a consequence,

there can be many solutions for a same n value. These states are said to be *degenerated*.

l is the azimutal quantum number and m is the magnetic quantum number. For the lowest values of the principal quantum number n, the following table gives the correspondences between the numbers (n, l, m) and the designation of orbitals:

$n = 1$ has only one state for $l = 0$ $(m = 0)$. It is the (1s) orbital.
$n = 2$ has only one state for $l = 0$ $(m = 0)$. It is the (2s) orbital.
$n = 2$ has 3 degenerate states for $l = 1$ $(m = -1, 0, +1)$. These are the (2p) orbitals.
$n = 3$ has only one state for $l = 0$ $(m = 0)$. It is the (3s) orbital.
etc.

Let us sort the orbitals in the increasing order of the pair (n, l) (Fig. 3): (1s)(2s)(2p)(3s). The corresponding series of the number of degenerate states is (1)(1)(3)(1). The following section use that series which is important when studying polyelectronic atoms.

5   Polyelectronic atoms

As we did for the hydrogen atom, it is possible to establish the equation of the mechanical energy for a polyelectronic atom (*cf.* Eq. (6)). Unfortunately, nobody has yet found, an analytical solution for atoms with more than one electron. It is interesting to make an analogy with the general solution for the trajectory of more than two bodies in their gravitational fields.

As analytic solutions do not exist, physicists try to model the electronic neighborhood of polyelectronic atoms by approximations.

In order to understand the electronic density of a polyelectronic atom, we use the chemist's approach. Chemists are used to sort atoms by families. For instance, it is possible to class them in increasing order of their atomic masses: (H, He, Li, Be, B, C, N, O, F, Ne, Na, Mg), etc. This series can be divided into sub-series according to the maximum number of hydrogen atoms that can be linked to them. As an example, the maximum association of hydrogen to the atoms (C, N, O, F) gives (CH_4, NH_3, H_2O, HF). The decreasing number of bounds with hydrogen atoms forms a subseries.

If we mix the increasing order of masses and the decreasing order of hydrogen associations, we obtain a series constituted by the sub-series (H, He)(Li, Be) (B, C, N, O, F, Ne)(Na, Mg), etc. This series can be symbolized by (2)(2)(6)(2) where the numbers correspond to the number of atoms in each sub-series. It is important to compare it with (1)(1)(3)(1) (*cf.* Sect. 4) corresponding to the number of degenerate states for each pair (n, l) for the hydrogen atom. These two series are related, linked by a factor two. The interpretation of this phenomenon leads to believe that polyelectronic atoms arrange their electrons with the same logic than the states of the hydrogen atom. Consider the oxygen atom as an example. It has 8 electrons. Place two of them in the (1s) orbital, two others in the (2s) orbital

H																	He
Li	Be											B	C	N	O	F	Ne
Na	Mg											Al	Si	P	S	Cl	Ar
K	Ca	Sc	Ti	V	Cr	Mn	Fe	Co	Ni	Cu	Zn	Ga	Ge	As	Se	Br	Kr

Fig. 6. The first four rows of the periodic table of the elements. The chemical series described in the text can easily be found.

and the last four in (2p) orbitals. The corresponding electronic configuration is $(1s)^2(2s)^2(2p)^4$.

An electronic configuration describes how electrons are placed in each type of orbital. For a given electronic configuration, there can be many combinations of electronic positions. Each combination is a state.

5.1 The spin

The chemists' approach leads us to postulate that the electron density of poly-electronic atoms should be not far from that of the orbitals of the hydrogen atom. We must explain where comes the origin of the factor two between the "chemists" series and the series of the degeneracy number of hydrogen orbitals. This factor can be explained if, at most, two electrons can occupy each orbital. Dirac theoretically demonstrated that each electron has a spin symbolized by an additional quantum number m_s which is equal to $-1/2$ or $+1/2$.

The fundamental origin of spin cannot be found in Equation (6). Spin comes from a rigorous treatment using relativistic quantum mechanics. It is impossible to describe it simplistically. The analogy of spin and the rotation of the electron is commonly used. However, it is a too simple an image that cannot explain the physical properties associated to spin.

In a given orbital, at most two electrons with opposite spins can be placed.

As a consequence, the orbitals and spins can be described as spinorbitals. A spinorbital is described by four quantum numbers (n, l, m, m_s). Spin induces few additional properties or complications. Firstly, the spin must respect the non-discernability principle. This principle stipulates that the exchange of two electrons in an atom does not change its properties. Let us have an illustration with the helium atom. In the ground state, its two electrons occupy the (1s) orbital. They have opposite spin. Let α be the symbol of the $m_s = +1/2$ spin function and β the symbol of the $m_s = -1/2$ spin function. Moreover, one of the two electrons is labelled 1 and the other is labelled 2. Electrons 1 and 2 of

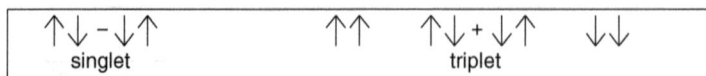

↑↓ - ↓↑
singlet

↑↑   ↑↓ + ↓↑   ↓↓
              triplet

Fig. 7. Symbolic notation of spin states for singlet and triplet states. The arrows' directions are used to distinguish α and β spin states.

ground-state helium must have their spin α and β, or β and α respectively. A mathematical notation could be $\alpha\beta + \beta\alpha$ or $\alpha\beta - \beta\alpha$. In the case of the $(1s)^2$ configuration of the helium atom, it can be demonstrated that only the $\alpha\beta - \beta\alpha$ solution is acceptable because it excludes the possibility of finding the two electrons at the same place. It is the singlet state.

The excited state $(1s)(2p)$ of the helium atom can also be a singlet state. However, as the two electrons do not occupy the same orbitals, they have no chance to be found at the same place. As a consequence, the $\alpha\beta + \beta\alpha$ function can also be found. The two other solutions $\alpha\alpha$ and $\beta\beta$ can also occur. In the case of helium atoms, these last three functions have almost the same energy and are called a triplet. The two functions $\alpha\alpha$ and $\beta\beta$ have parallel spins. The function $\alpha\beta + \beta\alpha$ has opposite spins. The triplet is often described as parallel spins. It is partially true.

The sum of the individual spins of electrons of an atom is noted S. The quantity $(2S+1)$ is usually called the spin multiplicity. We have studied the singlet $(S = 0)$ and triplet $(S = 1)$ multiplicities of the helium atom. For a given electronic configuration, many spin multiplicities are possible. For example, the $(1s)(2p)$ electronic configuration can be singlet or triplet (*cf.* Fig. 9).

It is important not to confuse the spin multiplicity and the line multiplets observed in spectra.

The line multiplets correspond to transitions between electronic states (*cf.* Sect. 7).

5.2   Hund's rule

For a given electronic configuration, the higher spin states are the more stable.

This remark is known by chemists as Hund's empirical rule. It is justified by nothing that the higher the multiplicity, the greater number of orbitals containing only a single electron. Electron-electron repulsions are reduced when electrons occupy orbitals singly. As a consequence, the higher the multiplicity, the greater the system's stability. The Hund rule is theoretically demonstrated because the energy of high-spin states is stabilized by a term called the exchange integral. The exchange integral has no equivalent in classical physics and it would be risky to try to give it a simple physical meaning.

5.3 Spin-orbit coupling

It is important to mention that the electronic configuration and the spin state are only approximations for polyelectronic atoms. This description is rather good for the helium atom and generally for atoms with few electrons. In the case of the helium atom, we said that the triplet spin state of the electronic configuration (1s)(2p) corresponds to three states of quasi equal energies. A detailed analysis shows that these three states are split by the fine-structure phenomenon. This comes from the electromagnetic interaction between the spin and the orbital angular momenta. The sum of these moments is noted J.

Spin-orbit coupling leads to many states with different energies for a given spin and electronic configuration.

As an example, the Figure 8 shows the lowest excited state of the sodium atom [Ne](3p) split into two states ($^2P_{3/2}$ and $^2P_{1/2}$) that are separated by an energy of 0.002 eV due to spin-orbit coupling. We will study that in detail in the next section.

Spin-orbit coupling leads to the existence of a very large number of energy levels. These levels are more numerous when orbital and spin angular momenta are highs. This is the case for the iron atom which has a quintet ground state (see Fig. 12).

5.4 Spectral terms and the levels

In order to simplify the notation, quantum numbers L and S are associated to each electronic configuration. L is the sum of the magnetic quantum numbers m of the occupied orbitals. S is the sum of each of the spin quantic numbers m_s. The spin-orbit coupling is taken into account by the quantum number J. It should be noticed that J can be any integer or half integer between $|L - S|$ and $|L + S|$, inclusive.

A spectral term is defined by the generic symbol $^{(2S+1)}L$ where S is the total spin number and L is the total azimutal number.

A spectral term has $(1 + 2$ times the minimum of $(L, S))$ levels from $J = |L - S|$ to $J = |L + S|$.

A level is defined by the generic symbol $^{(2S+1)}L_J$ where J is the total angular momentum.

A level is constituted by $2J + 1$ states which have the same energy in the absence of external perturbations. Each state of a level $^{(2S+1)}L_J$ is defined by the four quantum numbers (J, L, S, M) where M varies from $-J$ to $+J$.

Let us take the sodium atom whose ground-state electronic configuration is $(1s)^2(2s)^2(2p)^6(3s)^1$ as for ground state. The core electrons $(1s)^2(2s)^2(2p)^6$, have the same electronic configuration as the neon atom. To simplify the notation,

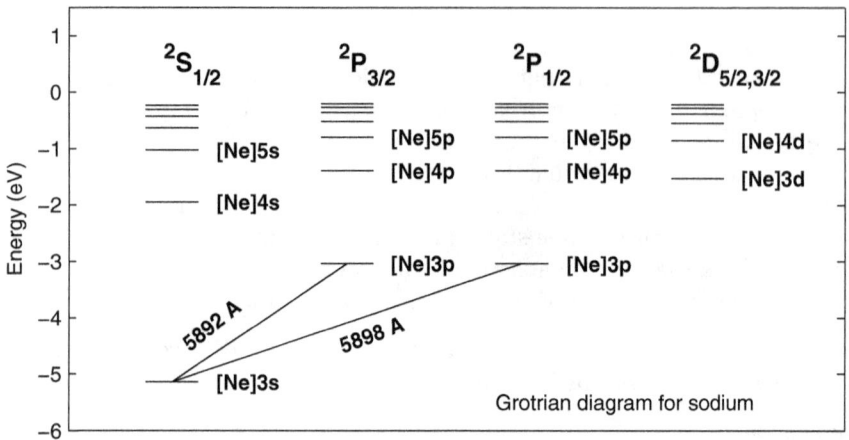

Fig. 8. The D doublet of sodium corresponds to two transitions observed at 5892 and 5898 Å from the ground level [Ne]3s to the electronic configuration [Ne]3p split by the spin-orbit interaction.

$(1s)^2(2s)^2(2p)^6(3s)^1$ is then replaced by $[Ne](3s)^1$ or by $[Ne](3s)$ because there is only one electron in the (3s) orbital.

The sodium [Ne](3s) electronic configuration is a doublet state ($2 = 2S + 1$ with $S = 1/2$). The (3s) orbital has an orbital azimutal momentum L. The $L = 0$ is then symbolized by S, $L = 1$ by P, etc. (notice the analogy with the l quantum number for the hydrogen atom). Hence the spectral term is ^2S. The spin-orbit coupling shows that $J = 1/2$ (J cannot be negative). There is only one level associated with the [Ne](3s) electronic configuration: $^2S_{1/2}$.

The excited state [Ne](3p) is also a doublet ($S = 1/2$). As the (3p) orbital has a total azimutal number $L = 1$, the spectral term is ^2P. The spin-orbit coupling leads to two possibilities: $J = 3/2$ and $J = 1/2$. The corresponding levels are $^2P_{3/2}$ and $^2P_{1/2}$. The sodium spectrum exhibits a doublet of lines at 5892 and 5898 Å corresponding to the transition from [Ne](3s) to the two levels of [Ne](3p). It is the famous D doublet observable on many celestial spectra.

The notations for levels $^{(2S+1)}L_J$, and those for the configurations $(1s)(2s)\ldots$, are complementary. For the helium atom, the $(1s)(2p)$ configuration leads to $L = m_{1s} + m_{2p} = 0 + 1 = 1$. The (2p) orbital has three m possible values $(+1, 0, -1)$ but the rule is to put electrons in the order of decreasing m values. Moreover, the $(1s)(2p)$ configuration can be singlet or multiplet (*cf.* Sect. 5.1). For the singlet ($S = 0$), possible value of J is $J = 1$ which leads to the level $^1P_1$. For the triplet ($S = 1$) the possible values of J are 0, 1 and 2 leading to the levels $^3P_0$, $^3P_1$ and $^3P_2$. The $(1s)(2p)$ configuration is then associated with four levels (hence four energy values): $^1P_1$, $^3P_0$, $^3P_1$ and $^3P_2$ (Fig. 9). As the same manner, it can be demonstrated that each configuration of type $(1s)(np)$, where $n \geq 2$, has four levels.

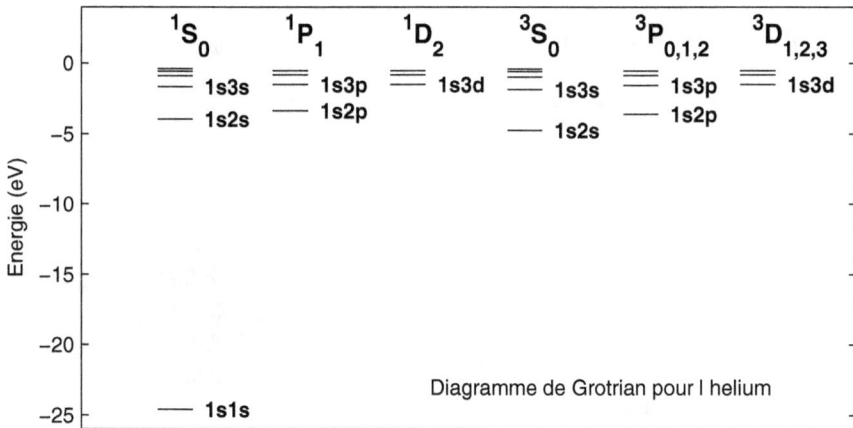

Fig. 9. Grotrian diagram of the helium atom for $L < 3$. The singlet states are also called para and the triplet states are also called ortho.

To avoid any ambiguity, each energy state of an atom should be designated by its electronic configuration AND its level (for instance, He $(1s)(2p)\,^3P_0$).

In a Grotrian diagram (Fig. 9) the states of the same level symbol are placed in the same column. It is a generalization of the Grotrian diagram of the hydrogen (Fig. 4) where columns are the same values of l.

Laboratory experiments show that the energy gap between different J levels of a given spectral term ^{2S+1}L increases as the nuclear charge increases. The electrostatic potential increases with the nuclear charges and the electrons closest to the nucleus have high velocities to avoid falling into the nucleus. For heavy atoms (iron for example) the velocity becomes close to that of the speed of light and one must take account of relativistic effects to be correct. The splitting of the levels increases with the relativistic effects. It is so small in light atoms that levels have quasi similar energies (but they are not strictly equal if the analysis is sufficiently detailed). For an helium atom, the levels $^3P_0$, $^3P_1$, $^3P_2$ are sometimes packed in the generic level $^3P_{0,1,2}$. It is an abusive notation but often used (see Fig. 9).

5.5 Tables of energy levels

The NIST (National Institute of Standards and Technology) has compiled energy levels for numerous atoms and ions measured since 1930. The web link is: http:// physics.nist.gov/cgi-bin/AtData/main_asd.

As an example, the Table 1 displays the lowest energy levels for an helium atom:

The Table 1 shows the very small energy difference between the levels $^3P_0$, $^3P_1$, $^3P_2$. This is the reason why they are packed in the generic level $^3P_{0,1,2}$ when high

Table 1. Extract from the NIST table for the first energy levels of the helium atom.

Configuration	Term	J	Level (cm^{-1})
1s2	^{1}S	0	0.00
1s2s	^{3}S	1	159856.0776
1s2s	^{1}S	0	166277.542
1s2p	^{3}P	2	169086.869782
		1	169086.946208
		0	169087.934120
1s2p	^{1}P	1	171135.00000
1s3s	^{3}S	1	183236.892
1s3s	^{1}S	0	184864.932
1s3p	^{3}P	2	185564.6651
		1	185564.6871
		0	185564.9577
	...		
He II	(^{2}S$_{1/2}$)	Limit	198310.7723

accuracy is not needed. The last line gives the limiting value beyond which the atom is ionized to He$^+$. This limit is the ionization energy of the helium atom. Its value is about 24.6 eV ($= 198310.7723/8065.5$).

The neutral helium atom is symbolized by HeI and the He$^+$ ion by HeII. The roman notations I, II, III usually designate the neutral, the first and the second stages of ionization. It is different to notations used for oxidation states used by chemists (0, I, II, ...).

6   Absorption and emission of energy

The absorption of energy by an atom is governed by the mechanical energy of states. The outcome is that an atom can only absorb energies that are exactly equal to the differences between two states. The absorption of energy leads to the excitation of the atom. The reverse phenomenon, the emission of energy, is also possible and leads to de-excitation.

There are two types of absorption process: Collisional and radiative.

The collisional process leads to the absorption of energy that comes from a moving object (atom, ion, *etc.*) that collides the atom. During the collision, part of the kinetic energy, corresponding to the difference between two states of energy of the atom, can be absorbed. A gas, at a given temperature, has motions with specific velocity properties. The collisions between atoms lead to energy exchanges. Some of the atoms remains in the ground state and the others are brought to excited states (see Sect. 9.1).

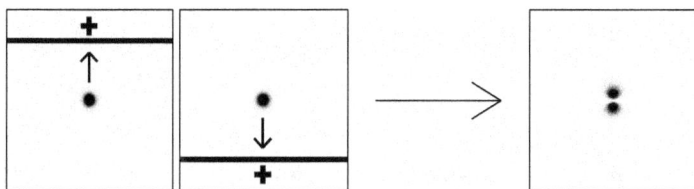

Fig. 10. *Left*: an orbital (1s) and an oscillating electric field (the arrow) due to the passage of the electromagnetic wave. *Right:* the oscillation of the field split the orbital (1s) in two parts and may put the electron in a (p)-type orbital.

In collisional processes, the higher the temperature, the more the atoms are brought to excited states.

6.1 Induced absorption

Radiative absorption is due to the interaction between the electromagnetic wave associated with the photon. It is induced absorption. In the range of optical wavelengths, only the electrons of the atoms can interact with the photons. The effects of the electric and magnetic waves of the photon on electrons must be considered. It can be demonstrated that the magnetic wave associated with the photon has no interaction. On the other hand, the electric field of the photon produces an electrical force on the electron of charge q from the classic law $F = q \cdot E$. Imagine an atom at the middle of this page and a photon that arrives from the left side. The wavelength associated with the optical light (about 5000 Å) is always much more larger than the dimension of the atom (a few Å). Consequently, the atom is submitted to an alternating electrical field with the frequency of the wave of the photon. The oscillation makes a force that goes up and down alternately. When absorption occurs, the electronic cloud is split into two parts, one up, one down leaving a void on the trajectory of the photon. Such a transition is called "electric dipolar" (usually noted E1). As a consequence, the final orbital is split into two parts. If the initial orbital is of (s) type, the final will be of (p) type.

It is demonstrated that the absorption of a photon in an electric dipole transition takes an electron from an orbital of azimutal quantum number l to values $l + 1$ or $l - 1$. This is the Laporte rule.

It is important to notice that a radiative absorption by an electron in a (1s) orbital cannot take an electron to orbitals (2s) or (3d). On the other hand, such an absorption can take an electron into (2p) or (5p) orbitals.

Besides the dipolar electric dipole transitions, there are also other type of interactions: electric quadrupole E2 ($\Delta l = 0$ or $l + 2$ or $l - 2$) and magnetic dipole M1 ($\Delta l = 0$). The intensities of these transitions are, in general, much smaller than those of electric dipole type. Nevertheless, they explain the emission of OIII oxygen by planetary nebulae at 5007 and 4959 Å (M1) and at 4367 Å (E2).

6.2   Oscillator strength and the Einstein coefficient

From the mathematical expressions of two final and initial wave functions, an efficiency coefficient can be calculated for absorption transitions. It is the oscillator strength (without any unit).

For an absorption transition, the oscillator strength (written f) is linked to the efficiency of the absorption. It is a decimal number with values between from 0 to 1.

The oscillator strength $f = 0$ corresponds to forbidden transitions and $f = 1$ to full efficiency. As the mathematical expression of the orbitals are approximations for polyelectronic atoms, the calculation of f is also an approximation. Many table values come usually from laboratory experiments.

A transition from a low level i of statistical weight g_i to a high level k of statistical weight g_k has $f = f_{ik}$. The statistical weight is defined by $g = 2J + 1$ and corresponds to the number of states in the level considered. In tables, one can find the Einstein coefficient for spontaneous emission, noted A_{ki} instead of f. They are related by the Ladenburg formula:

$$f_{ik} \cdot g_i = 1.499 \cdot 10^{-16} \, A_{ki} \, \lambda^2 \, g_k \qquad (9)$$

In formula (9), λ must be in Å and A_{ki} in s^{-1}. Tables give often A_{ki} in units of $10^8 \, s^{-1}$ or $\log(g_i \cdot f_{ik})$ without units (*cf.* Sect. 2).

6.3   Selection rules

Spin-orbit coupling induces a rule for the transitions that concerns the quantum number J. It can vary only by +1, 0 or −1 when a radiative transition occurs (except for $J = 0$ to $J = 0$, which is always forbidden).

It was said that for heavy atoms, the nuclear charge is high. They have such a large potential that internal electrons are relativistic. Consequently, relativistic effects induce perturbations that lead to new properties. For instance, transitions between different spins become allowed, although they are theoretically forbidden ($f = 0$) in the non-relativistic case.

The set of rules, that indicates which transitions are allowed, are usually called the selection rules.

Selection rules are used to identify allowed transitions from one level to another. In the case of electric dipole transitions: $\Delta l = +1, -1$ (Laporte rule), $\Delta J = +1, 0, -1$ (except for $J = 0$ to $J = 0$). $\Delta S = 0$ could be added for the lightest atoms.

6.4   Spontaneous emission

One of the inverse processes of absorption is spontaneous emission. When an atom is in an excited state, this state is unstable and it can emit a photon to come back to a more stable state. As there may be many energy levels that are more

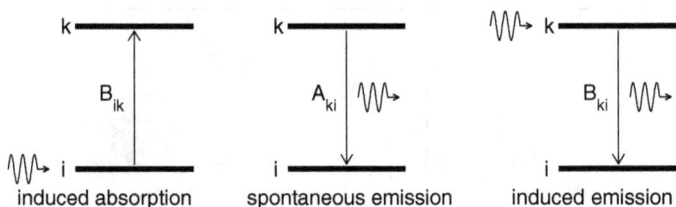

Fig. 11. Sketch of the three kinds of transitions between two energy levels i and k.

stable than the particular excited state involved, many de-excitation paths may be possible. The higher the value of f, the greater the probability of the path. The selection rules are the same as absorption.

Spontaneous emission is governed by the oscillator strength.

The efficiency of spontaneous emission is governed by the Einstein coefficient A_{ki} which is related to f by formula 9.

6.5 Induced emission

The second process of radiative emission comes from the interaction between an incident photon and the excited atom. This is induced emission. The efficiency of the induced emission is governed by the B_{ki} coefficient which is also related to the oscillator strength. There is competition between induced and spontaneous emissions.

7 Spectra

The spectrograph slit cuts through the studied object as a thin line. The optical set of the spectrograph projects the spectrum image formed by lines perpendicular to the spectral dispersion. These lines are the quasi monochromatic images of the entrance slit. Generally, experimental spectroscopists use the word "lines" for narrow spectral features, say less than a few Å. If spectral features are large or constituted by very close lines, it preferable to designate them as bands (the case of molecules). Theoretical spectroscopists use the words components, lines, multiplets to design spectral features and each of them has a precise sense.

In laboratories, atoms are often studied in their gaseous state at low pressure in a lamp (generally lower than a hundredth of atmospheric pressure). The lamp excitation is generated by electrical discharges which bring atoms into excited levels. These electrical excitations are not selective and every level can be populated. Each energy level of the atom is populated by a fraction of the atoms of the lamp.

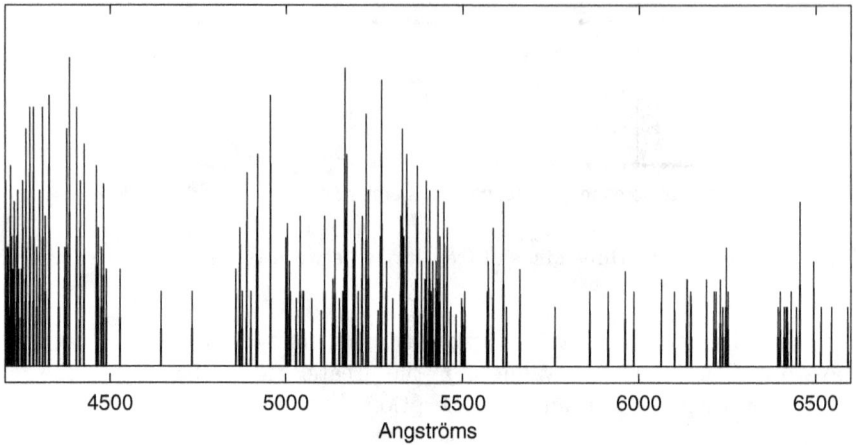

Fig. 12. Emission spectrum of the iron atom (FeI).

Excited atoms can undergo radiative transitions and emit photons with wavelengths that correspond to the energy difference between the level indeed.

Each transition produces an emission of light that is quasi-monochromatic. The total spectrum of the lamp is the sum of all the transition emissions. It is a discrete spectrum, like that presented in Figure 12.

Atoms can also undergo radiative transitions when they absorb photons whose wavelength corresponds to the energy differences between the levels concerned (this is induced emission).

A transition is a jump from one state to another. For a given initial level, there may be many final levels allowed for transitions. For the hydrogen atom, spectra obtained in the XIXth century allowed to be grouped the observed spectral features into series. For example, the Balmer series corresponds to the transitions from levels $n = 2$ to levels $n > 2$. The first spectral feature of a series is noted α, the second β, etc. For the Balmer series, Hα is for $n = 2$ to $n = 3$ transitions, Hβ is for $n = 2$ to $n = 4$ (see Fig. 14).

Due to the high complexity of spectral features, transitions are usually grouped with the following terms:

A line designates all the transitions between two different $^{2S+1}L_J$ levels.

A component is one of the transitions of a line.

A multiplet is the set of lines between two different ^{2S+1}L terms.

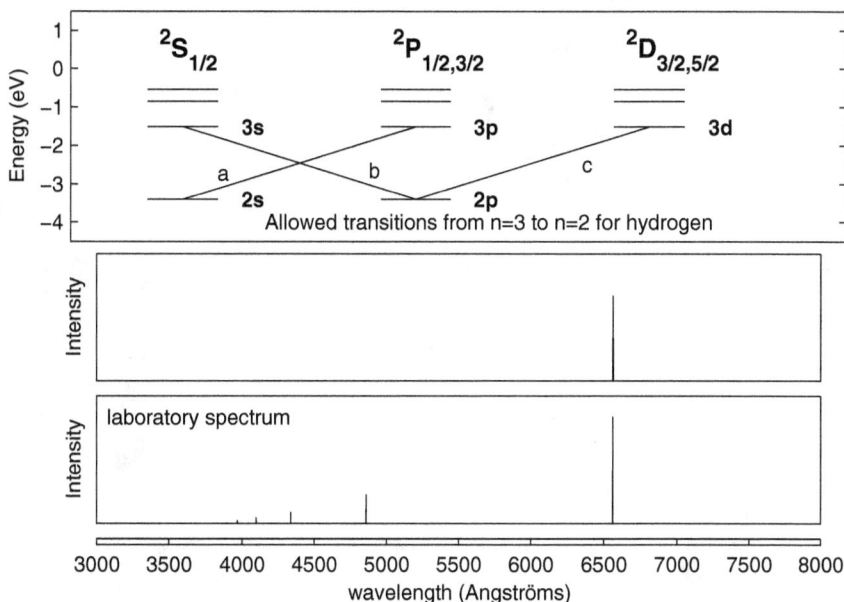

Fig. 13. *Top*: Grotrian diagram showing $n = 3$ to $n = 2$ transitions for hydrogen atom. *Middle*: the corresponding spectrum shows only one feature corresponding to Hα. *Bottom*: experimental spectrum shows Balmer series features.

The hydrogen atom spectrum obtained by very high resolution spectrographs shows that the "Hα line" designation is abusive. Due to spin-orbit coupling, Hα is constituted by three multiplets (labelled a, b, c in Fig. 13) with lines very close to each others, separated by less than 0.2 Å.

Consider the $(2s)^2S$-$(3p)^2P$ multiplet (labeled a on the Fig. 13). It is consti-tuted by a line doublet: $(2s)^2S_{1/2}$-$(3p)^2P_{3/2}$ and $(2s)^2S_{1/2}$-$(3p)^2P_{1/2}$ (the lowest energy level is always written first) at 6564.58 and 6564.54 Å respectively. It is equivalent to the multiplet of the sodium described in Section 5.4. The energy difference between the two lines is very small due to the very weak spin-orbit coupling in the hydrogen atom.

Consider the $(2s)^2P$-$(3s)^2S$ multiplet (labelled b in Fig. 13). For the same reasons, it is constituted by a line doublet at 6564.72 and 6564.56 Å.

Consider the $(2p)^2P$-$(3d)^2D$ multiplet (labelled c in Fig. 13). Applying the selection rules, it is constituted by a line triplet: $(2p)^2P_{3/2}$-$(3d)^2D_{5/2}$, $(2p)^2P_{3/2}$-$(3d)^2D_{3/2}$ and $(2p)^2S_{1/2}$-$(3d)^2D_{5/2}$ at 6564.66, 6564.68 and 6564.52 Å respectively.

Without any electric or magnetic perturbation, the line components are de-generate and Hα is constituted by only seven lines. In practice, the astrophysical conditions are such that lines are sufficiently broad that only one spectral feature is observed.

Submitted to an external electric or a magnetic field, each of the seven lines of Hα can be split into their components. For example, the $(2s)^2S_{1/2}$-$(3p)^2P_{3/2}$ line

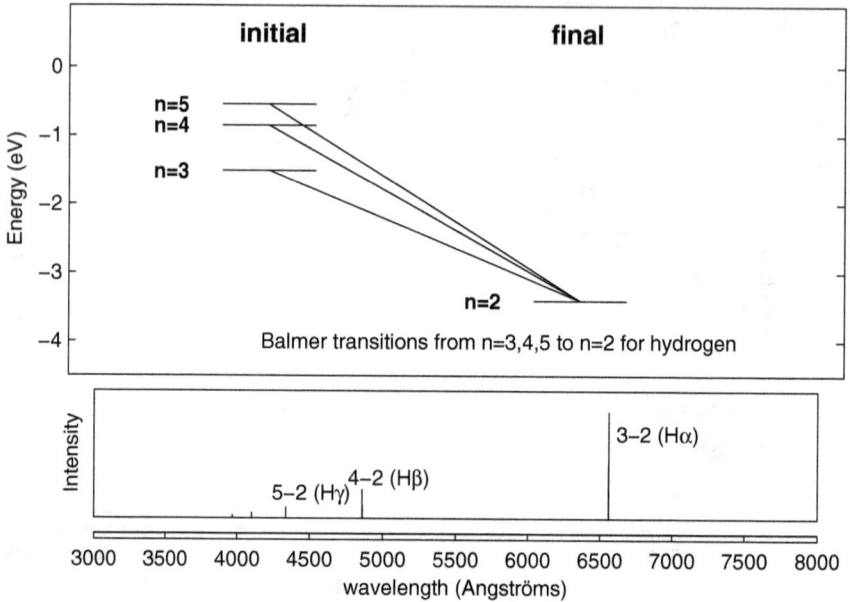

Fig. 14. Balmer transition from $n > 2$ to $n = 2$. Oscillator forces are $f_{H\alpha} = 0.64$, $f_{H\beta} = 0.12$, $f_{H\gamma} = 0.04$.

has six components. Remember that each level has $(2J + 1)$ states. Each state is associated to the four numbers (S, L, J, M) where M varies from $-J$ to $+J$. As a consequence, the $(2s)^2 S_{1/2}$ level is doubly degenerate and the $(3p)^2 P_{3/2}$ level is quadruply degenerate. As the selection rules for M are $\Delta M = -1, 0, +1$, six components are indeed expected. The seven lines have 48 components. Thus, $H\alpha$ corresponds to 48 allowed transitions in the hydrogen atom. These components can be observed when the gas lies in very intense magnetic fields.

In a spectral series, the oscillator strength decreases when difference in the principal quantum number increases.

This is due to the electric field effect on the initial orbital geometry. We have shown that the electric field of a photon can transform an (s) type orbital to a (p) one (Sect. 6.1, Fig. 10). For instance, the (2s) hydrogenic orbital will be transformed to (2p) "where the quote implicates that (2p)" has the same angular properties as a classical (p) orbital but the radial extension of the (2s). The oscillator strength will be high when the (2p)' can be superimposed on the final orbital. If we want to compare the oscillator strengths for (2s)-(3p) and (2s)-(4p), we must consider (2p)'-(3p) and (2p)'-(4p). The radial extension increases with n (cf. (2p) and (3p) in Fig. 3). As a consequence, (2p)' and (3p) can be superimposed better than (2p)' and (4p). We may generalize and conclude that the oscillator strength decreases when n increases in the series (2s) to (np).

For example, $f_{H_\alpha} > f_{H_\beta} > f_{H_\gamma} > \ldots$ As the intensity of a line is proportional to its oscillator strength, the intensity I tends to decrease in a spectral series: $I_{H_\alpha} > I_{H_\beta} > I_{H_\gamma} > \ldots$ This phenomenon is usually designed as decrement (Fig. 14).

The decrement of series can help in the identification of chemical elements in astrophysical spectra. Generally, the most intense transitions are those involving the lowest levels in a Grotrian diagram.

In a spectral series, the energetic differences of the highest levels becomes smaller and smaller when n increases. The energy of the highest levels tends asymptotically to the zero of mechanical energy, corresponding to the ionization of the species. Consequently, the associated transitions form a series of lines which become close together in the bluest part of spectrum, reaching the ionization limit as n tends to infinity (Fig. 14).

The hydrogen spectrum is constituted by many spectral series. The Lyman series is for the initial orbital (1s). The Lyα line is associated with the ground state of the hydrogen atom. All lines in the Lyman series occur in the ultraviolet range. All the Balmer lines ($n = 2$ initial level) lie in the visible range. The Paschen series ($n = 3$ initial level) lies partly in the visible range. The other lines and series are in the infrared and radio ranges.

The helium spectrum is more complex than that of hydrogen for two reasons. The first is that all the orbitals of the same n number do not have the same energy. The second is that helium can be in a singlet or in a triplet state. In the following study, to simplify, we consider the helium spectrum without any spin-orbit effects.

Consider an helium atom in its singlet state. The ground state is $(1s)^2$ the first excited state is $(1s)(2s)$. There is no allowed electric dipole transition between these two states because $\Delta l = 0$. The second excited level is $(1s)(2p)$. Then, the first radiative transition is $(1s)^2$ to $(1s)(2p)$. It is equivalent to the Lyα of the hydrogen. As for hydrogen, that series produces lines in the ultraviolet range.

It is interesting to study the equivalent of the Balmer series for helium. For the singlet state, there are three multiplets equivalent to Hα: $(1s)(2s)$-$(1s)(3p)$, $(1s)(2p)$-$(1s)(3s)$ and $(1s)(2p)$-$(1s)(3d)$. The same multiplets exist for the triplet state. Then there is a total of six multiplets to consider. As the intensity is highest for Hα, it is possible to predict, as a first approximation, the helium spectrum with these six multiplets (Fig. 15).

A single line cannot be used to identify a chemical element in a spectrum. Many lines are required and particularly those for the lowest energy transitions because these are the most intenses.

8 Line intensities

A medium is said to be optically thin if an atom's emission is not reabsorbed by the other atoms.

In optically thin media, emissions and absorptions are proportional to the number of atoms in the initial level. For a given level, it may exist many final

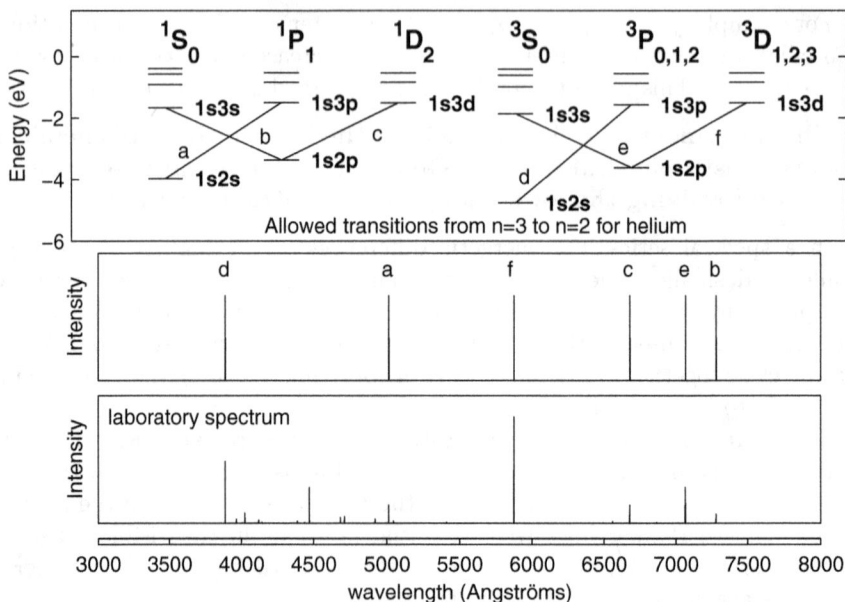

Fig. 15. *Top:* Grotrian diagram showing $n = 3$ to $n = 2$ transitions for helium atom. *Middle:* the corresponding spectrum shows six features corresponding to six possible multiplets. *Bottom:* the experimental spectrum shows that transitions between triplet states are more intenses than those between singlets.

levels. The final levels are accessed as their oscillator strengths or, equivalently, as their Einstein coefficients (*cf.* formula 9).

As a consequence of these principles, the energy emitted per unit of time, in any direction, by spontaneous emission line of wavelength λ from a high level (indexed k) to a low level (indexed i) is equal to:

$$E_{ki} = N_k A_{ki} \frac{hc}{\lambda} \tag{10}$$

where N_k is the number of atoms in the initial level. As time passes, N_k decreases as follows:

$$\frac{dN_k}{dt} = -N_k A_{ki} \tag{11}$$

If there is no excitation, N_k decreases to zero and the line stops emitting light. Emission lines observed in celestial objects are thus maintained to a constant N_k value by permanent excitation processes (collisions, induced absorption). In the case of a gas at thermodynamic equilibrium, the Boltzmann law (formula 13) described in the next section allows the N_k value to use in the formula 10 to be calculated, in order to compute the intensity of the line.

For induced absorption or emission lines, similar relationships are used adding the effect of energy brought by the photons.

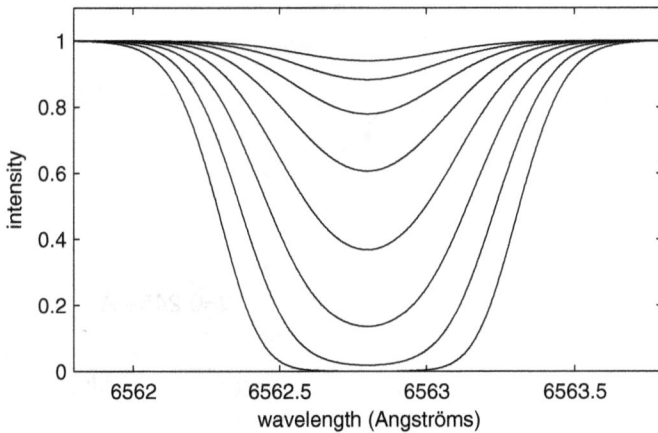

Fig. 16. A line broaden by a thermal Doppler effect (*cf.* Sect. 10) for various values of the optical depth. Only the least intense lines are optically thin.

In an optically thin medium, the intensity of lines is proportional to the number of atoms in the initial level and to the oscillator strength of the transition.

For optically thick media, growth of the line intensity is counterbalanced by self-absorption processes. It becomes difficult to estimate the N_k value in such media. A given element in a given medium can have both optically thin and thick lines. An optically thin medium must be rigorously associated with a given transition of an element (see Fig. 16).

8.1 The equivalent width

For a given transition, emission and absorption are quasi-monochromatic. Absolute monochromaticity does not exist and transitions are always broadened (*cf.* Sect. 10). From experimental spectra, the integrated energy is then measured over the range of the line-broadening.

In an experimental spectrum, the intensity is equal to the integrated area of the line. When the line is superimposed on the emission of a continuum, the equivalent width W is used.

The equivalent width is the width of a rectangle with the height of the continuum and an area equal to that of the line. Generally, the equivalent width is expressed in Å units. In an optically thin medium, the equivalent width is proportional to the number of atoms in the initial level, to the oscillator strength and to other parameters (the statistical weights of the levels).

For each observed line of a given element, W can be measured. The growth curve is a graph of W values versus parameters characteristic of the transition. In this way, abundance ratios of chemical elements can be determined in celestial

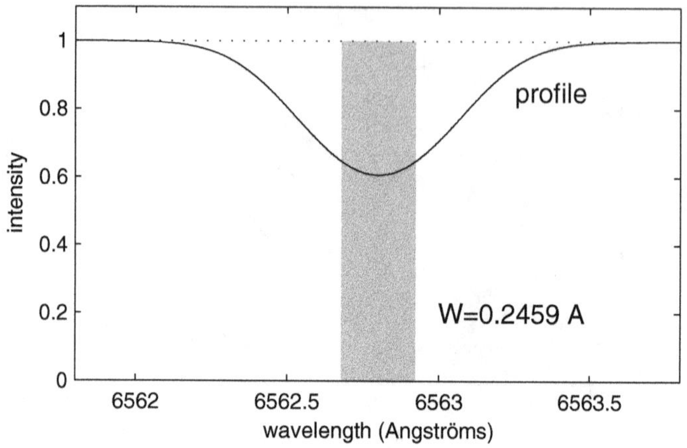

Fig. 17. The equivalent width is equal to the profile area of the feature under the continuum.

objects. Moreover, temporal variations of equivalent widths can reveal fundamental processes not detectable by classical photometry methods.

8.2 Transition tables

The NIST tables compiled for energy transitions of many atoms and ions (same web address as for the level energies tables). For instance, the following table is an extract of some transitions for the helium atom.

The second column of Table 2 gives the Einstein coefficient and the ninth and tenth columns the statistical weights. From them, the oscillator strengths can be calculated (from formula 9). It should be noticed that the wavelengths (first column) are quoted for standard air.

Table 2. Extract from the NIST table of some transitions of the helium atom.

Wavelength Air (Å)	A_{ki} (10^8 s^{-1})	E_i (cm^{-1})	E_k (cm^{-1})	Configurations	Terms	J_i	J_k	g_i	g_k
5875.9663	$3.92 \cdot 10^1$	169087.934120	186101.69622	1s.2p-1s.3d	3P*-3D	0	1	1	3
6678.1517	$6.38 \cdot 10^1$	171135.00000	186105.06984	1s.2p-1s.3d	1P*-1D	1	2	3	5
6867.48									
7065.179	$1.54 \cdot 10^1$	169086.869782	183236.892	1s.2p-1s.3s	3P*-3S	2	1	5	3
7065.217	$9.25 \cdot 10^2$	169086.946208	183236.892	1s.2p-1s.3s	3P*-3S	1	1	3	3
7065.710	$3.08 \cdot 10^2$	169087.934120	183236.892	1s.2p-1s.3s	3P*-3S	0	1	1	3
7281.351	$1.81 \cdot 10^1$	171135.00000	184864.932	1s.2p-1s.3s	1P*-1S	1	0	3	1
7816.15									

9   Populations

Even in astrophysical media where the gas is extremely diluted, it can be considered that volumes studied are so large that they contain large numbers of atoms. Models are generally based on volume sliced in layers, where gas conditions (pressure, temperature, radiations) are supposed constant. In these elementary volumes, it is important to calculate the atom fractions which lie in excited levels. This is the population analysis.

An energy level of an electronic configuration is said to be populated when there are atoms, inside a gas layer, that are in this state.

Theoretical calculations of population of electronic levels are particularly suitable for the spectral analysis. As inputs, there is a volume filled by a number of atoms at a given temperature and other factors such as radiations from external sources (other layers, a nearby star or the interstellar radiation field of our Galaxy). From these data, the theoretical populations of energy levels are computed and then a theoretical spectrum can be produced. The task of the astrophysicist is to find a model that gives a good agreement between the theoretical and the observed spectra. The population analysis is a very important step in spectral analysis.

Considering a large number of atoms excited only by collisions, the population of each level depends essentially on the gas temperature and the atoms' ionization energy. After a long time, the equilibrium is governed by the Maxwell-Boltzmann theory (Eq. (13)). This is called the thermalisation.

9.1   Saha and Boltzmann laws

One of the basic models is to consider that the gas is at thermodynamic equilibrium. That approach allows a simple law to be used to compute the population of the various levels. Two formulas should be used: the first to compute the population energy level ratios of an atom (Boltzmann law) and the second allows the ratio of neutral and ionized atoms to be computed (Saha law).

The Saha law states that the number of ionized atoms depends on the physical parameters temperature (T) and the electronic density (N_e). The electronic density is the number of electrons in a unit of volume. At low temperature, the ionization increases, because the gas temperature is linked to its velocity (collisional absorptions). At high temperature, the population decreases because neutralization of the ions. As a consequence, there is a range of temperatures where ionization reaches a maximum.

$$\frac{N^+}{N^0} = \frac{A(kT)^{3/2}}{N_e} e^{-\frac{X}{kT}}. \tag{12}$$

In the Saha formula (Eq. (12)), the A parameter is a function of the temperature, of statistical weights and the energies of electronic states. The X parameter is the ionization potential and k is the Boltzmann constant.

The Boltzmann law shows that high energy levels' populations increase as the gas temperature.

$$\frac{N_k}{N_i} = \frac{g_k}{g_i} e^{-\frac{E_i - E_k}{kT}}.$$ (13)

The indices i and k are associated to E_i and E_k respectively, and the g factors are the statistical weights (the number of degenerate states $g = 2J + 1$).

It is useful to couple the Saha and Boltzmann laws to predict the line intensities in the case of thermondynamical equilibrium.

As an illustration, we calculate the case of the Hα hydrogen intensity obs-break erved in absorption in stellar atmospheres. In these atmospheres, the most significant parameter is the temperature. We choose three cases: 5800 K (the Sun), 10 000 K (Vega) and 15 000 K. The Boltzmann law shows that $n = 2$ of the hydrogen atom is always slightly populated compared to the $n = 1$ level. As a consequence, computation is limited to the two lowest levels. The following table shows the population of a hydrogen volume containing about one billion atoms. Table 3 shows that the ionized atoms (column H$^+$) increases with temperature. The number atoms in the $n = 2$ state increases with temperature up to 10 000 K. Beyond this limit, the H($n = 2$)/H($n = 1$) ratio increases again but as abundance of neutral hydrogen becomes very low, the absolute population H($n = 2$) decreases.

Table 3. Populations of hydrogen against the temperature.

T	H($n = 1$)	H($n = 2$)	H$^+$
5800 K	$\approx 1.0 \cdot 10^9$	5	750
10 000 K	$\approx 0.5 \cdot 10^9$	9000	$0.5 \cdot 10^9$
15 000 K	$\approx 0.05 \cdot 10^9$	1000	$0.95 \cdot 10^9$

Since the line intensity is proportional to the number of atoms in the initial state, the Hα line ($n = 2$ to $n = 3$) should be proportional to H($n = 2$). As a consequence, the Hα line intensity should rise to a maximum near T 10 000 K and then decrease, due to the high hydrogen ionization (Fig. 19).

Another illustration of the Saha and Boltzmann laws concerns the intensities of the so called H and K calcium lines in the solar spectrum. These lines are much more intense than the Hα of hydrogen although calcium is less abundant by a factor of a million (see Fig. 20).

If one considers a volume containing about one billion atoms, it contains about 2000 calcium atoms. Calcium has a lower ionization energy ($X \sim 6$ eV) than hydrogen ($X \sim 13.6$ eV) and is essentially ionized in the solar photosphere. Populations are given for $T = 5800$ K in Table 4.

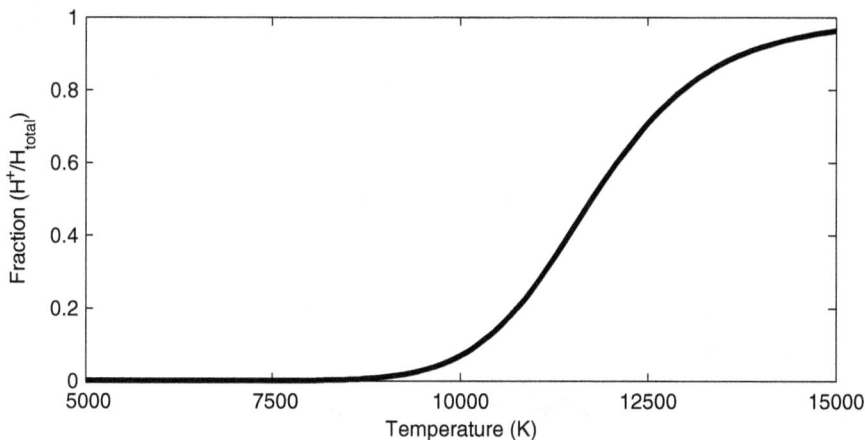

Fig. 18. Fraction H^+/H_{total} *versus* surface temperature of the star (according to the Saha law with $A/N_e = 1.03 \cdot 10^{25}$ cm^3). Ions become dominant above 12 000 Kelvin. As a consequence, the Hα line decreases beyond this limit because neutral atoms are rarely found.

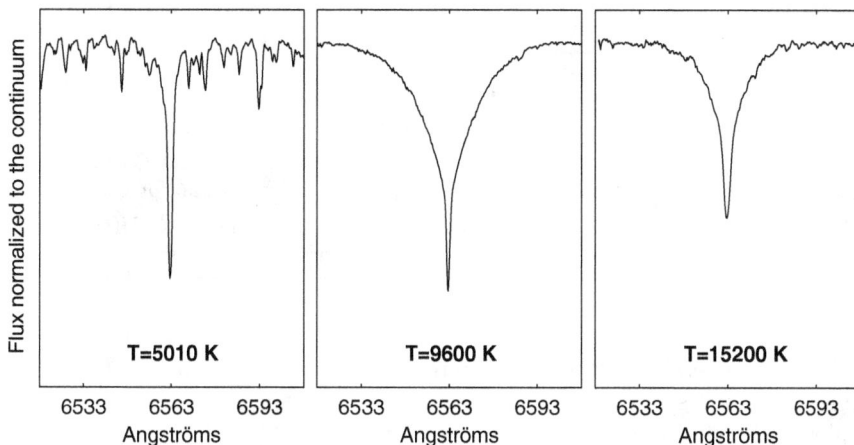

Fig. 19. Hα profile evolution for three stars with increasing surface temperature (kindly provided by Christian Buil).

Table 4 shows that the initial state population for Hα is 5, although it is 2000 for the initial state for CaII(H and K). This implies that the H and K lines of ionized calcium are much more intense than Hα in the solar spectrum (although calcium is about one million times less abundant than hydrogen). The calculation predicts similar intensities for the CaI(2) line of neutral calcium and for Hα, as it is observed.

The materials of our Galaxy are cycled from stellar formation clouds to stars and *vice versa*. The more massive stars have life times much less than the age of

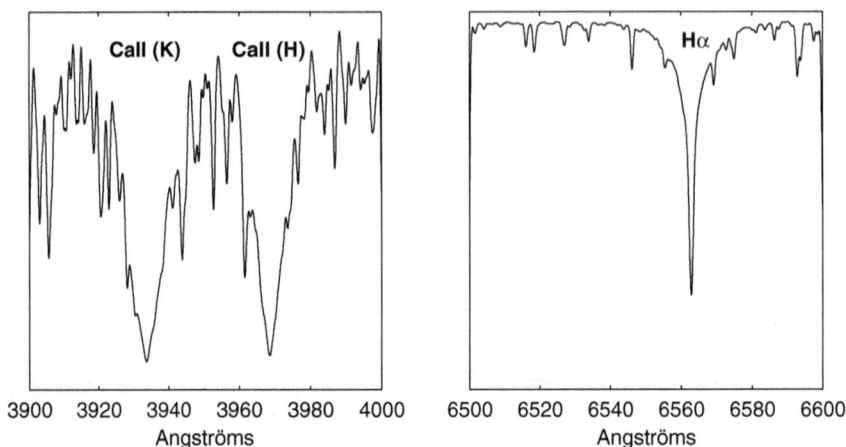

Fig. 20. Solar spectra showing the H and K calcium lines to be much more intense than the Hα line.

Table 4. Populations of some hydrogen and calcium levels for T = 5800 K.

Elements	State	Population	Comments
H	$(n = 1)$	$1.0 \cdot 10^9$	H ground state
H	$(n = 2)$	5	H First excited level. Initial for Hα
H	$(n = 3)$	< 1	H Second excited level. Final for Hα
Ca	ground	2	Ground state Ca. Initial for CaI(2)
Ca	1st excited	< 1	First excited Ca level. Final for CaI(2)
Ca^+	ground	2000	Ground state for Ca^+. Initial for CaII(H and K)
Ca^+	1st excited	7	Ca^+ first excited level.

the Galaxy which leads to homogeneity in the relative abundances of the chemical elements. In general, analysis of line intensites is undertaken not to measure the atomic abundances in celestial bodies, but to constrain acceptable values of the gas physicial conditions.

10   Line profiles

Experiments show that lines are always broadened by many processes. Broadening is very useful because it contains much information about the physical conditions of the medium. Helped by very-high resolution spectrometers (say $\lambda/\Delta\lambda > 20\,000$), studies of line profiles have enabled many astrophysical problems to be elucidated.

10.1   Natural broadening

The natural broadening of a line is a consequence of the uncertainty principle. We have already remarked (Sect. 2.1) that it is not possible to know both the

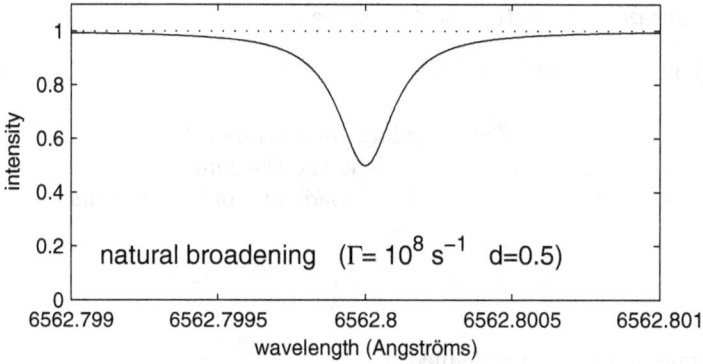

Fig. 21. Lorentzian profile due to the natural broadening of a line.

position and the velocity of a nanometer particle at the same time. One of the consequences concerns the life-time of excited states. Let us take the case of spontaneous emission. From an excited level, an atom can de-excite by transitions toward lower levels. The most probable transitions are those with a higher value of the oscillator strength. For common allowed transitions, the life time is about 10^{-8} s. The uncertainty principle stipulates that the product of the life time and the wavelength range of the radiation is equal to a constant, which leads to a broadening of about 0.0001 Å in the typical cases.

This natural broadening cannot be eliminated. It leads to the following expression for line shapes (Lorenzian function).

$$\frac{I(\nu)}{I(c)} = 1 - \frac{d}{4\left(\frac{\nu-\nu_0}{w}\right)^2 + 1} \tag{14}$$

where $I(\nu)/I(c)$ is the intensity compared to the continuum. d is the depth of the line, ν_0 the central frequency and w the characteristic width of the Lorentzian. For natural broadening, w is defined by:

$$w = \frac{\Gamma}{2\pi} \tag{15}$$

where Γ is a function of Einstein coefficients of the atom. The full width at half maximum $(FWHM)$ is related to w for a line centered at λ by:

$$FWHM = \frac{w\lambda^2}{c}. \tag{16}$$

Spectrographs must have very high resolution to measure the natural broadening.

Natural broadening of lines cannot be measured in the celestial spectra because many other factors contribute to much larger broadenings.

10.2   Effects of temperature and pressure

This section is devoted to broadenings due to physical conditions of the medium.

When the gas is inside a given volume at a temperature T, atoms move in any direction with variable velocities. This is the thermal turbulence, usually called by thermal Doppler. This effect adds a broadening of the line which is a Gaussian function:

$$\frac{I(\nu)}{I(c)} = 1 - d \cdot exp\left(-\left(\frac{\nu - \nu_0}{w}\right)^2 ln(16)\right). \tag{17}$$

For a thermal Doppler, w is defined by:

$$w = \sqrt{\frac{2kT}{m}}\frac{\nu_0}{c}\sqrt{ln(16)} \tag{18}$$

where m is the atomic mass, T the temperature of the gas and c the velocity of light.

In stellar conditions, temperatures are about 10 000 K. This leads to a thermal Doppler profile of about 0.5 Å. Generally, this effect hides the natural broadening.

In dense media (A, B, O type stars), if the mean time between two collisions is lower than the life-time of the excited states, transitions will be broadened by the Stark effect. This effect is due to the interaction between electric fields in the neighborhood of the electrons. The corresponding profile is of Holtsmark type, which is close to a Lorentzian with a Γ parameter defined by:

$$\Gamma = \frac{2}{t_c} \tag{19}$$

here, t_c is the mean time between collisions defined by:

$$t_c = \frac{1}{n \cdot D}\sqrt{\frac{m}{16\pi kT}} \tag{20}$$

where n is the density of the gas, D the atomic diameter (about 1 Å), m the atomic mass and T the temperature. For stars with a surface temperature higher than 10 000 K, the broadening is larger than 10 Å. This explains why the Balmer lines are very large in the spectrum of Vega (type A0V, T $\sim$ 10 000 K in Figure 19 middle panel).

Photospheric lines of stars are generally broadened by a thermal Doppler in the line core and by a Lorentzian profile in the wings.

The final profile of a line is the convolution of a Dirac by the Lorentzian and Gaussian profiles that includes the physical conditions (Thermal Doppler, Stark, etc.). Convolution by Gaussian and by Lorentzians lead to Voigt profile (pronounced *focht*). The *spectrum* program allows us to simulate a stellar spectrum

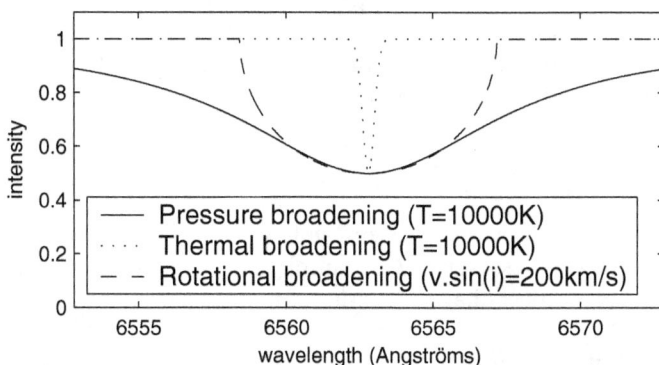

Fig. 22. Broadening of a line due to various physical processes.

from a set of physical conditions as inputs. It is a very pedagogical piece of software. It allows one to vary the parameters and shows the consequences on the spectrum. The URL is: `http://www.acs.appstate.edu/dept/physics/spectrum/spectrum.html`

10.3 Effect of magnetic fields

A magnetic field can also contribute to the modification of the line profiles. The consequence is a splitting into many components around the initial wavelength. This is the Zeeman effect. Let us take the Hα line (*cf.* Sect. 5.4). In usual conditions, splitting is rather low and can be seen only as a broadening of the line that is less than 1 Å. Nevertheless, photons of the line are polarized and measurement of the line polarization enables the magnetic field parameters to be determined. This is a very delicate procedure reserved for the best spectrographs.

The energy of a state defined by its J quantum number is modified by $g\mu_B JB$, where B is the intensity of the magnetic field, μ_B is the Bohr magneton and g is the Landé factor which is about equal to:

$$g \sim 1 + \frac{J(J+1) + S(S+1) - L(L+1)}{2J(J+1)}. \tag{21}$$

The Landé factor shows that the splitting increases with the spin multiplicity but decreases with the value of L. For instance, in magnetic field studies of the sun spots, the line $^5P_1$-$^5D_0$ of FeI at 6173.3 Å is often used. The levels have a high spin multiplicity ($S = 2$) and a low value of L. The Landé factor for the level $^5P_1$ is $5/2$.

10.4 Shifts and broadenings by Doppler-Fizeau

The previous broadening factors are due to the physical conditions of the medium: temperature, density, mass, electronic density. Other line-shape modifications are due to dynamical conditions of the medium. This is the Doppler-Fizeau effect. It can be demonstrated that the observed wavelength λ is shifted from its theoretical value λ_0. In the case of non-relativistic velocities, the difference $(\lambda - \lambda_0)$ can be related to the ratio of the velocity v and the velocity of light c.

$$\frac{\lambda - \lambda_0}{\lambda_0} = \frac{v}{c}. \tag{22}$$

This relation shows that wavelengths are shifted cowards the blue when the object moves towards to the observer. When an object moves away, the shift is to the red. In our Galaxy, objects have velocities of about $\pm 10^4$ m/s. These imply shifts of about 1 Å.

At this level, it is important to notice that Earth orbital motions (around the Sun and around its axis) add a telluric Doppler-Fizeau component to the observed spectra. It is important to correct the wavelengths for these effects. This is a barycentrical correction. When the objects are external to the solar system, it is necessary to subtract the velocity of the Sun with respect to the nearest stars. The set of the corrections allows us to obtain wavelengths placed in the local standard of rest (L.S.R.).

The Doppler-Fizeau effect can also produce broadenings in the case of objects that rotate. When an observer is placed in the equatorial plane of rotation of an object, he sees an edge that comes cowards and the opposite edge that goes away. As a consequence, the composition of every Doppler-Fizeau effect, from one edge to the other, leads to a broadening of the lines. Hot stars generally have high speeds of rotation, and this effect becomes important. The rotational broadening effect cannot be detected if the object is seen from one of its poles. It can be demonstrated that the broadening measurement allows one to determine the value of the product $v_e \cdot sin(i)$ where v_e is the equatorial rotation velocity and i is the pole inclination against the line of sight. The convolution function of a spectrum by a rotation profile is:

$$\frac{I(\lambda)}{I(c)} = 1 - d \cdot \sqrt{1 - \left(\frac{\lambda - \lambda_0}{v_e \cdot sin(i)} \right)^2}. \tag{23}$$

In the case of emission by an expanding shell (planetary nebulae for instance), the observer measures a broadening of lines because one face comes back and the other face goes away. The profile due to an expanding shell is different to that of rotation, and the two phenomena can be well separated.

10.5 Instrumental broadening

Instrumental broadening comes from the physical laws that limit the spectral resolution achieved by a spectrograph. For instance, the total length L of a diffraction grating acts as a window in the sense of physical optics. It produces a broadening proportional to the inverse of L. This explain why high spectral resolution can be obtained only by using large (and expensive!) gratings. Moreover, the photon captor is made up of microcells (pixels in the case of a CCD captor) which also limit the resolution. The profile generated by an ideally monochromatic source is called the instrumental function, and written PSF for Point Spread Function.

Instrumental functions produce a spectral broadening which is mainly due to the optical elements of spectrographs.

11 Continuum

The difference of energy between the ground state and the zero-level of mechanical energy is the potential energy necessary to remove one of its electrons from an atom. It is called the ionization potential. All levels of bound electrons lie in the range of the ionization potential. The number of states lying in a slice of energy can be counted. This number is called the density of states. The Grotrian diagram of hydrogen shows that levels becomes closer together when the energy increases. This is true whatever the atom considered. As a consequence, the density of states increases with the energy. At the limit of the zero mechanical energy (ionization of the atom), the density of states tends to be infinite and it is the continuum. In that case, the electron is free, because its mechanical energy is positive and can reach any value.

If many atoms are associated in a small volume, the density of states of the material increases proportionally to the number of atoms. Electronic interactions between atoms must be added to describe the matter. They are mainly electrostatics and lead to modify slightly the energy levels. The simplest case arise for molecules. Molecular spectra are much more complex than those of atoms because the number of levels are increased due, for instance, to vibrations of bonds. The line series for molecules can be so rich that they form large bands.

If the atom number is sufficiently high, only the density of states can be computed. This is the case of solid materials and for very dense gases which constitute the stellar mater just under the photosphere. The emission can no longer be described as jumps between separated levels, and combinations are so numerous that the matter emits at every wavelength. The problem can be solved by statistical models.

11.1 Black body laws (Wien and Stefan)

If a material, that emits continuous radiation can also absorb photons of any wavelength, it is called a black body. This is an idealized notion which is a good approximation for many celestial bodes (stars, planets, dust grains, etc.). When it emits its own light, radiation from a black body is governed by specific laws demonstrated by Planck at the beginning of the XXth century.

The emission properties of a black body are not linked to the nature of its components but only to its surface temperature.

The radiation emission of a black body reaches a maximum for a wavelength λ_m. The Wien law (Eq. (24)) shows that the product of λ_m and the temperature T must be constant:

$$\lambda_m \cdot T = 2.88 \cdot 10^{-3} \, mK. \qquad (24)$$

The bolometric luminosity is the radiation emitted for all the wavelengths in every direction. The bolometric luminosity of a black body is proportional to the fourth power of its surface temperature. This is the Stefan's law (Eq. (25)). That law explains immediately why the hottest stars are those with the shortest life-time. The Sun has a total life-time estimated to be about 10^{10} years for a surface temperature of about 6000 K. If its temperature will doubled, it would emit $2^4 = 16$ times more radiation power. If everything else equal, its lifetime should be 16 times less, say a few hundreds of thousand years. This calculation is very simple and one must add other processes to be more realistic. However, it shows the real tendency. The Stefan law formula is:

$$M = \sigma \cdot T^4 \quad (\text{units: W/m}^2) \qquad (25)$$

where σ is the Stefan-Boltzmann constant ($\sigma = 5.67 \cdot 10^{-8}$ international system units). M is the emittance of the black body, that is the power emitted by a surface unit (M is expressed in W/m^2).

The Wien and Stefan laws are basic tools for astrophysicists because they allow them compute important parameters (temperature, energies, etc.) from the continuum of observed spectra.

In practice, classical stars exhibit a continuum emission close to that of a black body. This is almost the total energy radiated by the star. The emission occurs as soon as the upper layers of gas become sufficiently diluted not to reabsorb the totality of photons. The layer which emits the continuum is called the photosphere and absorption lines in the stellar spectra are usually found in the layers immediately above it.

11.2   Non thermic continuous radiations

Other physical processes can also generate continuous spectra. The ejection of an electron from an atom requires an energy at least equal to that of the ionization potential. But any higher energy can also ionize. This is a bound-free absorption. As a consequence, the ionization spectrum is an absorption continuum which begins at the energy of ionization. The most typical case is from the hydride anion H^- in cold stars. Its absorption is so strong that the continuum of type M stars is affected by it in the whole range of visible and near infrared wavelengths. Another case is the discontinuity of Balmer.

Another continuous emission can result from a loss of kinetic energy by free electrons slowed down by electrostatic effects. This is free-free emission, also called Bremsstrahlung. This process is observed in many ionized gaseous nebulae. Other continuous emissions can be observed in very special conditions, such as the synchrotron radiation which results from the interaction between charged particles and a strong magnetic field. It is observed in some quasars, around neutron stars, supernovae, etc.

The set of continuous emissions generated by other processes than the black body radiations are called non thermal emissions.

12   The spectral analysis

The spectral analysis of a spectrum can be sketched as follow. The continuum allows the temperature to be determined in the case of black-body emission. Many lines for many elements allow the chemical components of the gas to be identified. Equivalent-width measurements allow chemical abundances in the gas phase to be calculated. Shifts and profile analysis allow pressure and dynamical effects to be determined.

To conclude, it is appropriate to build a model which can predict the same spectrum as that observed. This includes the knowledge of the geometrical description of the source possibly sliced into many layers. The goal is to compute all of the parameters for each layer. The spectral analysis is an investigation where spectral features are the clues. It is the daily work of astrophysicists.

I thank Valerie Desnoux and Maylis Lavayssiere for their helpful remarks after reading the first manuscript. I thank also professor Colin Marsden of the Toulouse University for time he spent for english corrections.

Astronomical Spectrography for Amateurs
J.-P. Rozelot and C. Neiner (eds)
EAS Publications Series, **47** *(2011) 39–71*

SPECTROGRAPHS IN AMATEUR ASTRONOMY

C. Buil[1]

(Translation by Stephen Dearden)

Abstract. This paper reviews the overall principles of astronomical spectrographs and the design rules that cover low, medium, and high spectral resolution instruments. Several examples of spectrograph designs are described that are easy to build and are optimised for amateur telescopes. Results are given for each of these instruments. Despite the modest size of amateur telescopes, we show that high performance spectrographs in the hands of amateur astronomers provide access to some specialised areas where the amateur could contribute to useful astrophysical work.

1   Introduction

The usual way to analyze light from a celestial object spectroscopically is to split a light ray into its component wavelengths and record the weak signals obtained with an electronic detector. Since modern electronic detectors are array detectors, one pixel will receive for example blue light while another pixel further away will receive red light. The number of pixels per wavelength unit will define the level of dispersion, which is itself linked to the capability to separate and distinguish small details in the spectrum.

In a sense, the dispersion process is in contradiction with traditional astronomical imaging. In conventional imaging, the objects observed are point light sources and the criterion for good image quality is the concentration of the light onto the smallest possible number of pixels at the focal plane of the telescope. In spectroscopy, the same information is deliberately spread over a large number of pixels. Since each pixel receives only a portion of the total photon flux, it is necessary to use a highly sensitive light sensor, capable of extracting the faint

---

[1] 6 place Clémence Isaure, 31320 Castanet-Tolosan, France; e-mail: `christian.buil@wanadoo.fr`; `http://www.astrosurf.com/buil`

© EAS, EDP Sciences 2011
DOI: 10.1051/eas/1147002

signal from the noise background which is inherent to each spectral data acquisition. It goes without saying that the other components of the optical system being used must not be neglected. Using a telescope with the largest possible mirror to collect the maximum number of photons in a minimum length of time will of course help enormously. The overall transmittance of the complete instrument chain should be carefully optimized in order to minimize the loss of photons owing to the atmosphere at the detector. Characteristics such as the sensitivity of the sensor, the telescope's light gathering capacity, the transmission factors of the optical elements and the quality of the atmosphere, will all determine the final overall performance of the spectrograph. This term is also known as the *throughput*.

Spectroscopy is therefore quite a demanding discipline, requiring the use of high throughput optical systems. This is one of the main reasons why spectral analysis was for a long time out of reach of the budgets of amateur astronomers, and remained the domain of professional astronomers with large telescopes and sophisticated electronic technology. However, since the latter half of the 1990's, the situation for amateurs has vastly improved. Amateur astronomers are now routinely using large format CCD (Charge Coupled Device) electronic cameras which have the right performance characteristics in terms of photon and wavelength sensitivity that are well adapted to the requirements for spectroscopy. Today, the commercial cameras available for amateurs compare very favourably with those employed by the professional when it comes to comparing readout noise, the most important performance parameter for spectroscopy.

The problem of telescope diameter, however, still remains. The majority of amateurs have telescopes that range from 0.2 meter to 0.6 meter in size. This is relatively small when compared to the giant eyes employed by the professional astronomer. Nevertheless, the key to success in spectroscopy lies in the choice of the instrument design concept which can actually benefit greatly from the small size of an amateur telescope.

Leaving aside telescope diameter for the moment, it is perfectly possible to construct, at relatively low cost, a spectrograph that has a throughput equal to, or better than, a professional version. One can even show that for the same spectral resolution, the physical size of the spectrograph, which essentially determines its final manufacturing cost, is directly proportional to the diameter of the telescope. Thus an amateur using a 200 mm telescope can quite easily construct a spectrograph with a resolving power equivalent to that employed on giant telescopes at world-class observatories. Of course, the difference in diameter between professional and amateur telescopes still has a very significant impact, the main factor being that far fewer stars are accessible to amateurs. But by making the right instrumental choices, the number of possible targets is still at a level where some excellent, and publishable, scientific work in specific areas can be performed. Some of these applications will be described later in this article.

The technical choices to be considered first will be for spectrographs directly mounted at the telescope's focal plane. Configurations using optical fibers or light guides to transfer photons from the telescope to a spectrograph located some distance from the instrument will not be considered in this article.

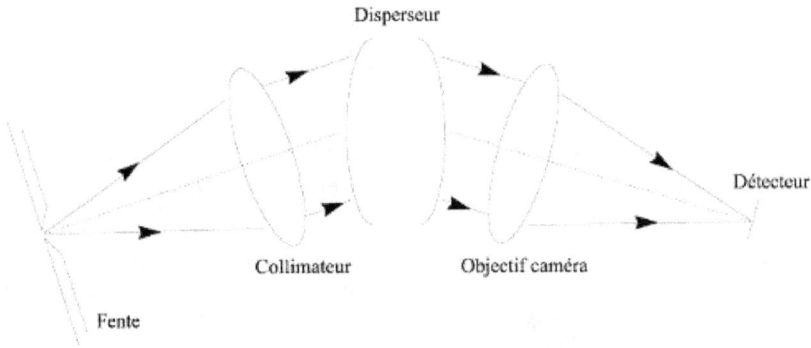

Fig. 1. Generalized schematic diagram of a spectrograph.

Nevertheless, the equations presented here are system independent and are still relevant.

When we physically attach the spectrograph to the telescope, the total weight at the telescope becomes an important factor. An additional load of 2 to 3 kg is usually the maximum limit for most amateur telescopes. Often, less expensive solutions have to be considered, such as using standard photographic lenses for the construction of the spectrograph. The design should be as simple as possible in order for the amateur to build himself, which is a further way of reducing cost.

After the basic design principles of spectrographs are described, this chapter continues with the basic definitions and design criteria needed to meet the specifications given above. The formulae will allow the reader to easily define the various characteristics of a particular spectrograph configuration that best suit his or her needs. Three examples of spectrograph configurations that cover the definitions of low, medium and high spectral resolution will be presented as an illustration of the design rules explained in this chapter.

2 Basic principles of spectrographs

Spectrographs based on light dispersion devices are all built on the same principle, regardless of whether the instrument is required for a backyard telescope or for a large profession telescope such as the VLT. Light rays from a source object interact with the following five key components (see Fig. 1):

1. A narrow *slit*, which serves as the entrance aperture of the spectrograph, and which is located at the focal plane of the telescope. One of the functions of the slit is to isolate a very narrow and defined part of the field image.

2. A lens (or mirror), which converts the diverging light beam from the slit into a parallel beam. This optical component is the *collimator lens* or more simply the *collimator*.

3. A light *dispersion* device, whose role is to separate (disperse) polychromatic light rays from the source object into its components wavelengths.

4. A *camera* lens which forms a focused spectrum on the plane of the detector. More accurately, it is an image of the slit that is focused at the detector and there are as many images of the slit as there are wavelengths in the analysed light.

5. A *detector* (*e.g.* CCD camera, digital SLR). This will record the flux distribution as a function of wavelength that represents the spectrum.

The function of the collimator is to minimize optical aberrations of the dispersing device. Once the optical beam is collimated, light rays coming from a given point on the slit fall with the same angle of incidence angle on the disperser. Consequently, for any given wavelength, the rays are deviated by exactly the amount, finally converging at the same point on the detector.

The two usual dispersion devices employed are (**a**) the prism and (**b**) the diffraction grating. The prism was the mainstay disperser for all spectrographs prior to the invention of the grating, it is a simple optical component and small prisms are inexpensive. However, prisms have several drawbacks when it comes to building a spectrograph.

The diffraction grating is in many ways much better suited for modern applications. With a grating, the dispersion process is based on the constructive and destructive interference effects that occur when light rays pass through a structure consisting of a very large number of narrowly spaced, parallel and equidistant engraved lines or grooves. The many varied colours seen on the surface of a Compact Disc (CD), when you look at a CD close to grazing incidence, is an example of light dispersion by a grating. In a fully functioning spectrograph, the diffraction grating is generally a plane glass substrate on which a very large number of grooves have been ruled or, more commonly these days, etched, using a specialised industrial process, to a high degree of precision. If the grating is a reflection grating, the lines are made within a thin layer of aluminium deposited on glass substrate in order to reflect light. The number density m of the grooves or lines per unit length provides information on the thickness of the etching. A grating mechanically ruled with a diamond head can achieve up to 1,800 lines per millimetre (lpm). Holographic etching techniques allow gratings to be produced with groove densities of 3,200 lpm and even 4,800 lpm. The higher the density, the higher the dispersion power of the grating. This relationship allows for much flexibility for different applications.

As an illustration, let α be the angle of incidence on the grating of a light ray of wavelength λ. The direction of diffraction β with respect to the grating normal[1] is derived from the so called *grating formula*:

$$\sin \alpha + \sin \beta = k \, m \, \lambda. \tag{1}$$

In this equation, a large deviation of a light ray corresponds to a high value of m. The amount of deviation is also dependent of the parameter k, an integer which

---

[1]The normal is the perpendicular to the grating surface at the impact point of the light ray. The incidence and direction angles are counted in the same plane. These angles are of opposite sign if they are located on each side of the normal.

can take values of ..., -2, -1, 0, $+1$, $+2$, ... Different values of k correspond to what are termed different grating *orders*, and each order exhibits its own distinct spectrum. A diffraction grating, in fact, produces multiple spectra simultaneously. When $k = 0$, the grating behaves like a simple mirror ($\alpha = -\beta$). In general, an amateur spectrograph will only exploit the 1st order spectrum (where $k = 1$ or $k = -1$).

By modifying the shape of the grating grooves to give a saw-tooth profile (a technique known as *blazing*) it is possible to concentrate a larger amount of the light (up to 70%) into a single grating order. The blazing technique can also be used to maximise grating efficiency at a particular wavelength within a given order. The grooves form facets from which the angle is modified and which act as many very tiny mirrors. Such a grating is called a *blazed* grating and many modern grating used today are blazed. As light photons are scarce in astronomy, a blazed diffraction grating is preferred when a grating has to be selected from a manufacturer's catalogue. A grating blazed for 1st order around a wavelength of 500 nm is a good choice in most cases. However, it has to be understood that the grating efficiency will gradually fall from this maximum value. Such gratings can readily be found on the market at a reasonable price[2].

A fundamental characteristic of a spectrograph is its capability of separating two closely spaced wavelengths. This is analogous to the angular resolving power of a telescope to separate, for example, the two components of a double star. Spectral sharpness, $\Delta\lambda$, is defined as the finest spectral detail separated at the wavelength λ. It is usual in spectroscopy to evaluate the quantity R, a dimensionless quantity, called the *resolving power* or *resolution*, which immediately leads us to an expression for the resolution of the instrument:

$$R = \frac{\lambda}{\Delta\lambda}. \tag{2}$$

Spectrographs with values of $R < 500$ are considered to be a low resolution instruments and the spectra such instruments record are low resolution spectra. At values of R between 500 and 5,000 the spectrograph is known as a medium resolution device and when $R > 5,000$ we start to enter the high resolution range.

Let w be the physical width of the entrance slit of the spectrograph at the focal plane of the telescope. The angular width of the slit ω, in radians, as seen looking from the entrance of the telescope, is given by

$$\omega = \frac{w}{f} \tag{3}$$

---

[2]To give some idea, a 50 mm × 50 mm grating costs about 200 Euros (2005 prices). The following websites can be consulted for further details: http://www.edmundoptics.com, http://www.optometrics.com, http://www.gratinglab.com.

where f is the focal length of the telescope. ω is also called the projected slit width on the sky[3]. The expression for the resolving power, which is a function of ω, is

$$R = \frac{\lambda k \, m \, d_1}{\omega D \cos \alpha} \tag{4}$$

where D is the telescope diameter and d_1 the usable diameter of the collimator lens. The proof of this important equation will be given later, but for now let us examine some of its properties.

Consider first the ratio d_1/D. This ratio is telling us that in order to maintain the same spectral resolution as the telescope diameter increases, we must increase all dimensions of the spectrograph in proportion. This is one of the fundamental reasons for the great potential of spectrographs in the hands of amateurs. For a large professional telescope it is necessary to design a huge, expensive and heavy spectrograph, whereas to achieve a roughly equivalent performance result the requirements for an amateur design attached to a modest telescope are much less demanding.

It is very clear, of course, that the quantity of light photons collected (the photon flux) will be far higher with a professional scope and this is important in an overall assessment of performance. Nevertheless, since there are quite a number of spectroscopically interesting objects of magnitudes down to 6–8, amateurs can make a significant contribution in this discipline. The narrowness of the slit (small ω) also plays an important role in increasing the resolution. Some technical limits do remain, the principal problem being that too narrow a slit width will greatly reduce the flux entering the spectrograph. In fact, due to the atmospheric turbulence, a star's image at the focal plane of the telescope is not strictly a point source. If the aparent angular width of the star, defined by the *seeing* angle ϕ, exceeds the angular width of the slit, the quantity of light entering the spectrograph rapidly diminishes. In practice, from a sea-level site, it is not practical to work with a projected slit width smaller lower than 3 arc seconds, thus $\omega = 1,45 \times 10^{-5}$ rd.

For some instrument configurations and applications the spectrograph can be used without a slit, which confers a significant advantage from the point of view of efficiency and ease-of-use. However, this working in this so-called *slitless* mode is a special case. Slitless spectroscopy is considered in the description of the first spectrograph in the following section with the MERIS spectrograph, a medium resolution instrument.

3   MERIS: A medium resolution spectrograph

The acronym MERIS stands for Medium Resolution Imaging Spectrograph. Used with a modest refractor of, say, 100 mm diameter, MERIS can record spectra down to magnitude 9 with a signal-to-noise ratio (S/N) of 100 and an exposure

---

[3]In case of the use of an optical fiber, the quantity ω is the diameter of the fibre core projected on the sky.

time of one hour. The number of objects accessible with such a combination is greater than 100,000! MERIS uses standard photographic lenses, relatively easy to find on the second-hand market at reasonable prices these days and at Internet web sites such as eBay. On the other hand, the amateur may already have this type of lens among his or her old photographic equipment. The collimator is a 135 mm telephoto lens and the camera objective lens is a standard 50 mm f/1.8 SLR lens (the faster the camera objective lens the better – see later in this section).

The grating is a reflection grating, 30 mm in width with 600 lines per millimetre, blazed in first order at a wavelength of 500 nm[4]. This choice will be justified later on in this paper. The CCD camera is equipped with a sensor very well known to the amateur astronomy community, namely the Kodak KAF-0402ME chip with a 768 × 512 array of 9 micron pixels.

3.1 Geometric parameters

Let us first consider the spectrograph mounted at the focal plane of a $D = 200$ mm diameter telescope with a focal length f of 1200 mm. If F_t is the focal ratio between the focal length and the aperture, $F_t = f/D = 1200/200 = 6$. In order to avoid any optical vignetting of the light beam transmitted by the telescope, the beam passing through a collimator lens, whose useful relative aperture is, say, F_c, the following condition must be met:

$$F_c < F_t. \tag{5}$$

Since the large majority of 135 mm telephoto lenses have focal ratios of f/2.8 or faster, this criterion is easily achieved. The effective focal ratio used is F_c' so that $F_c' = F_t$. The diameter d_1 of the light beam exiting the collimator is thus given by

$$d_1 = \frac{f_1}{F_t} = D \frac{f_1}{f}. \tag{6}$$

In our example,

$$d_1 = \frac{f_1}{F_t} = \frac{135 \ (\mathrm{mm})}{6} = 22.5 \ \mathrm{mm}.$$

The useable width, L, of the grating, in the dispersion direction, i.e. perpendicular to the grooves and in the plane of the grating, is derived from the following equation (see also Fig. 2)

$$L = \frac{d_1}{\cos \alpha}. \tag{7}$$

Substituting d_1 from Equation (6) in Equation (7) we obtain

$$L = \frac{f_1}{F_t \cos \alpha}. \tag{8}$$

---

[4]Reference 45343 in the Edmund Optics catalogue.

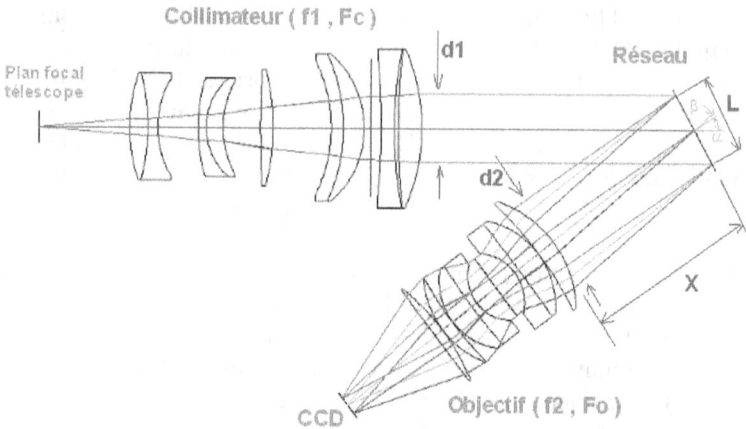

Fig. 2. MERIS spectrograph. The lens of the collimator is shown here only schematically. The proportions are representative of typical photographic lenses.

Since diffraction grating costs increase rapidly with physical size, it is important to calculate the minimum acceptable grating size required and then optimise the various other parameters that depend upon it. The formula in Equation (8) shows that if we decrease the focal distance f_1 of the collimator lens, we are going in the right direction. However, additional considerations described below with regard to spectral resolution suggest that the above choice of $f_1 = 135$ mm is close to the optimum for a MERIS type spectrograph. Since the focal ratio F_t is linked to the telescope being used, the only variable that can be changed in the end in order to control the grating size is the angle of incidence α of the light beam.

The grating formula (1) shows that there is an infinite number of values of α associated with the angle β, as long as the term on the right hand side of Equation (1) is met. In practice, the choice is much more restricted due to physical and mechanical reasons. Let the quantity γ, called the *total angle*, be defined as:

$$\gamma = \alpha - \beta. \tag{9}$$

The total angle is defined in the plane of incidence between an incident ray and a diffracted ray at a wavelength λ_0 (refer to Fig. 2). The wavelength λ_0 is known as the *centered wavelength* and, as its name implies, is found at the centre of the spectrum that will ultimately be recorded.

If γ is too large, the physical size of the spectrograph becomes unnecessarily large and the value of the angle α (or β) is such that the size of the grating is too large (expensive!) or the speed (focal ratio) of the camera lens needed to accept the beam from the grating is too wide, and technically difficult to achieve at reasonable cost. Conversely, if γ is too small, the barrels of the collimator and objective lenses get in the way of each other and the light beam is cut off. Note that moving the grating further from the front lens of the camera (distance X in

Fig. 2) is no solution and must be absolutely avoided at all costs, since this would drastically reduce the overall throughput of the spectrograph, as we shall see later. A maximum typical value for X is between 50 and 60 mm.

In practice, with standard photographic lenses, we quickly reach the following trade-off: $\gamma = 38°$ and $X = 60$ mm. These values are in fact the ones adopted for MERIS. In order to be convinced of this argument, the reader can set up the same configuration on a PC using optical ray tracing or CAD software, or you can even draw the setup the old fashioned way with paper and pencil, following the spectrograph schematic described above. When copying in this way, care must be taken to ensure that the optical axes of the collimator and camera lenses join at the grating plane centre. Once these mechanical criteria are set, the two values of α and β, at the wavelength λ_0, are pre-determined. By introducing the angle γ into the grating formula, we can establish the relation

$$\alpha = \arcsin\left(\frac{k\,m\,\lambda_0}{2\,\cos\left(\gamma/2\right)}\right) + \frac{\gamma}{2}. \tag{10}$$

For each diffraction order, $+1$ and -1, there exists a distinct value for α. If we take $\lambda_0 = 5500\,\text{Å}$, which is the approximate centre of the visible spectrum, with the parameters for a MERIS type spectrograph, we have

$$\alpha = \arcsin\left(\frac{600 \times 0.55 \times 10^{-3}}{2 \times \cos\left(38/2\right)}\right) + \frac{38}{2} \simeq 29.05°$$

and

$$\alpha = \arcsin\left(\frac{-600 \times 0.55 \times 10^{-3}}{2 \times \cos\left(38/2\right)}\right) + \frac{38}{2} \simeq 8.95°.$$

The corresponding diffraction angles are derived from the formula (9) giving, respectively $\beta = -8.95°$ and $\beta = -29.05°$.

Which is the best choice for β? In most cases, the best solution is the one which has the grating pointing perpendicular to the camera lens as opposed to the collimator lens. This design rule offers better linearity of dispersion and maximizes the quantity of diffracted photons. The final choice for MERIS is therefore $\beta = -8.95°$. The minimum size L of the grating can now be calculated from Equation (7) to give

$$L = \frac{22.5}{\cos(29.05°)} \simeq 25,7 \text{ mm}.$$

The grating dimensions are quite realistic since the etched surface is $30 \times 30\,\text{mm}^2$. However, the margin of error is fairly small along the direction of the plane of incidence – only 2 mm on each side of the light beam. It is therefore necessary to align the optical axis very carefully and to prevent any mechanical flexure. After being diffracted, the light then goes to the camera lens. At the grating output, the beam width d_2 measured in the plane of incidence is given by

$$d_2 = L\,\cos\beta = d_1\,\frac{\cos\beta}{\cos\alpha} = \frac{f_1}{F_t}\,\frac{\cos\beta}{\cos\alpha}. \tag{11}$$

The numerical value here is $d_2 = 25.7 \times \cos(8.95°) \simeq 25.4\,\text{mm}$. The beam width is, however, significantly greater at the entrance to the camera lens due to the spectral dispersion effect. At the distance X of the grating this dimension is approximately equal to

$$d_2' = \frac{f_1}{F_t}\frac{\cos\beta}{\cos\alpha} + \frac{X\,p_\lambda\,N_x}{f_2} \tag{12}$$

where f_2 is the focal length of the camera lens, N_x is the number of CCD pixels along the dispersion direction and p_λ the size of a pixel in this same direction. The first term of Equation (12) is nothing more than Equation (11). The additional second term includes the angular dispersion and the width of the final recorded spectrum. With a KAF-0402 sensor we have $N_x = 768$ and $p_\lambda = 9\,\text{microns}$, thus

$$d_2' = \frac{135}{6}\frac{\cos 8.95°}{\cos 29.05°} + \frac{60 \times 0.009 \times 768}{50} \simeq 33.7\,\text{mm}.$$

The diameter of the entrance pupil of a 50 mm focal length lens at $f/1.8$ is equal to $50/1.8 = 27.8\,\text{mm}$. This aperture is not large enough to collect all the photons and bring them to focus across the full length of the spectrum. Only the central region of the spectrum experiences the full field irradiance. At the edges of the spectrum vignetting becomes significant and this can be as high as 20%, although this is still a reasonable level for a spectrograph of this type. In order to minimize these losses the fastest possible camera lens should be employed and opened to full aperture. An $f/1.4$ lens is a good choice and within the range of amateur budgets if acquired second-hand from Internet sites such as eBay.

The angular dispersion $d\beta/d\lambda$ at the grating output is expressed in radians per Ångström unit[5], and is given by

$$A = \frac{d\beta}{d\lambda} = 10^{-7}\frac{k\,m}{\cos\beta}. \tag{13}$$

The linear dispersion l in the detector plane is the product of the angular dispersion and the focal distance of the camera lens, thus

$$l = \frac{dx}{d\lambda} = f_2\frac{d\beta}{d\lambda} = 10^{-7}\frac{f_2\,k\,m}{\cos\beta}. \tag{14}$$

Let us now evaluate the formula in the neighbourhood of the centre wavelength, i.e. when $\beta = -8.95°$:

$$l = 10^{-7}\frac{50 \times 600}{\cos 8.95°} \simeq 3.04 \times 10^{-3}\,\text{mm/A} = 3.04\,\text{microns/A}.$$

The inverse of the linear dispersion P (sometimes called the *plate factor*[6]), expressed in Å per millimetre, is given by

$$P = 10^7\frac{\cos\beta}{k\,m\,f_2} \tag{15}$$

---

[5] We will use the Ångström unit (Å) that is still in regular use in astronomical spectroscopy although the standard SI unit is the nanometre (nm). $1\,\text{nm} = 10\,\text{A} = 10^{-9}\,\text{m}$.

[6] From the time when photographic plates were used as the detector in spectrographs.

and the inverse dispersion in Ångströms per pixel is

$$\rho = 10^7 \, \frac{p_\lambda \, \cos\beta}{k \, m \, f_2}. \tag{16}$$

For 9 microns pixels around the centre wavelength of 5500 Å, this calculation produces

$$\rho = 10^7 \, \frac{0.009 \times \cos 8.95°}{600 \times 50} \simeq 2.96 \ \text{A/pixel}.$$

Thus a pixel covers a spectral range of about 2.96 Å.

On each part of the centre wavelength the detector records a spectrum limited on wavelength by λ_1 and λ_2 so that

$$\lambda_{1,2} = \lambda_0 \pm \frac{\rho \, N_x}{2}. \tag{17}$$

With $\lambda_0 = 5500$ Å, $N_x = 768$ pixels and $\rho = 2.96$ Å/pixel, the limits are $\lambda_1 = 4494$ Å and $\lambda_2 = 6506$ Å. It can be seen that only a relatively limited region of the spectrum is recordable with a solid state sensor such as a CCD due to the limited number of pixels. To record other parts of the spectrum, from the infrared around 1 micron (1000 nm or 10,000 Å) to the near ultraviolet around 350–400 nm (3,500 Å to 4000 Å), the only solution is to change the centre wavelength. This is achieved most practically by rotating the grating around an axis parallel to the grooves. It is quite straightforward to change the central wavelength and still maintain a constant angle γ. The alternative approach would be to move the camera across the grating surface to scan the spectrum, but this is an unnecessary, and mechanically more difficult.

3.2 Spectral resolution

Consider two spectral lines of equal intensity at the wavelengths λ and $\lambda + \Delta\lambda$. We usually measure the Full Width at Half Maximum (FWHM) of the line, δ_t. There is a simple criterion for determining spectral resolution, known as the Houston criterion, that states that two adjacent spectral lines are considered separated if their centres distance are greater than or equal to the quantity δ_t. The spectral *finesse* (or sharpness), $\Delta\lambda$, is similarly defined as the smallest perceptible detail, lying on a linear dimension δ_t in the focal plane of the camera lens. From the spectral dispersion formula (15), we therefore obtain the following expression

$$\Delta\lambda = \frac{\delta_t \, \cos\beta}{k \, m \, f_2}. \tag{18}$$

Now the resolving power (or resolution), R, of the spectrograph was defined in Equation (4) at the beginning of this chapter. The FWHM for a given observed line, δ_t, will depend on the physical width of the entrance slit (or the diameter of an optical fibre) and in the way that different optical components of the spectrograph

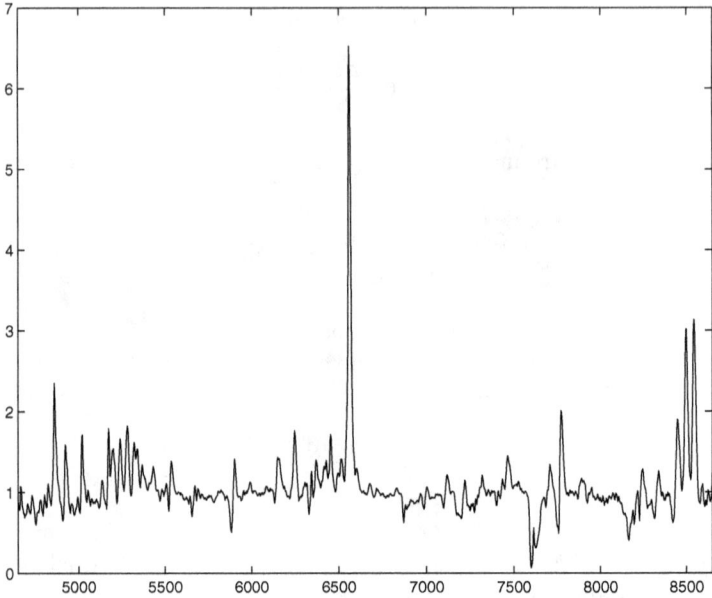

Fig. 3. A typical spectrum taken with the MERIS spectrograph: Nova Scuti 2003 observed on September 3rd 2004 at magnitude $V = 8.7$. The total integration time was 30 min. with a 128 mm refractor. The Hα emission line is near the centre.

form this image on the detector. In general, unwanted diffraction effects and optical defects in the collimator and camera lenses blur the image of the slit and thus degrade the spectral resolution finally achievable. If δ_d, δ_c and δ_o represent this degradation in quality slit image due to diffraction, collimator lens aberrations and camera lens aberrations respectively, the effective apparent width of the slit can be expressed approximately as[7]

$$\delta_t = \sqrt{\left(\frac{r\,f_2}{f_1}\right)^2 (w^2 + \delta_c^2) + \delta_o^2 + \delta_d^2} \tag{19}$$

where r is known as the anamorphic coefficient and is equal to $r = \cos\alpha/\cos\beta = d_1/d_2$. As long as $\delta_t > p_\lambda$ the spectrum is optimally sampled since at least two pixels cover a resolution element (Nyquist criterion). The resolving power can then be written as:

$$R = \frac{\lambda\,k\,m\,f_2}{\delta_t \cos\beta}. \tag{20}$$

---

[7]Equation (19) is approximate because we give here the quadratic sum of the contributing factors. This will produce an exact result only if these contributors are independent of each other and if the blur function exhibits a Gaussian distribution. This is not rigorously correct, but the error is small enough as not to affect the result significantly.

If $\delta_t < 2\,p_\lambda$ the spectrum is under-sampled, the resolution is pixel-limited, and the resolving power is expressed by

$$R_{pixel} = \frac{\lambda\,k\,m\,f_2}{2\,p_\lambda \cos\beta}. \tag{21}$$

When the entrance slit aperture is larger than the seeing disk of the star, the term w in Equation (19) must be replaced by the linear dimension u of its stellar image at the telescope's focal plane. The relevant expression is given by $u = \phi\,f$. Additionally, if u is the dominant term in Equation (19), the spectral resolution is referred to as *seeing-limited*. In this case, the spectral resolution equation becomes

$$\triangle\lambda = \frac{\phi\,\cos\alpha}{k\,m}\,\frac{f}{f_1} \tag{22}$$

and the resolving power can be written as

$$R_{seeing} = \frac{\lambda\,k\,m\,f_1}{\phi\,f\,\cos\alpha}. \tag{23}$$

Under these conditions, a large slit width, or even no slit at all, can be used, thereby considerable improving the photometric throughput of the spectrograph because all the light flux from the star passes directly into the spectrograph without any obstruction. Used in this way it is a relatively straightforward task to position the target star correctly at the centre of the field of view (as seen in the CCD image) and maintain the star there during a full integration time. This is to be compared with attempting to keep a target star image positioned between the jaws of the slit, a dimension often not more than a few tens of micrometres, which is a much more challenging task when performed manually.

As astronomers all know, seeing conditions are certainly not constant over time and vary appreciably. Since it is not a very satisfying approach to allow the vagaries of the Earth's atmosphere to restrict the quality and reproducibility of data collected, it is most important that the spectrograph limits this potential variable spectral spreading due to the atmospheric turbulence. For example, Equation (23) shows the advantage of using a short focal length telescope (f), which is most often the type found at amateur observatories. The same conclusion can be made for telescope diameter D, since Equation (23) can be re-written as

$$R_{seeing} = \frac{\lambda\,k\,m\,d_1}{\phi\,D\,\cos\alpha}. \tag{24}$$

For a given spectrograph, the smaller the diameter of the telescope, the higher the spectral resolution. We can also draw the conclusion, as already mentioned, that to maintain the spectral resolution constant, if the telescope diameter grows, the spectrograph must also increase in physical size accordingly. To illustrate this concept, let us calculate the resolving power of MERIS under 3 arcseconds seeing conditions that represents a typical value for an amateur site. The seeing angle is first converted into radians using the formula

$$\phi_{radian} \simeq \frac{\phi_{arcsec}}{206265} \tag{25}$$

with $\phi = 3'' = 1.45 \times 10^{-5}$ rd.

If the telescope diameter is 200 mm and a wavelength of 5500 Å $(0,55\times10^{-3}$ mm), we obtain, from Equation (24)

$$R_{seeing} = \frac{0.55 \times 10^{-3} \times 600 \times 22.5}{1.45 \times 10^{-5} \times 200 \times \cos 29.05°} \simeq 2900.$$

The linear dimension δ_s of the seeing disk translated onto the plane of the detector is thus given by

$$\delta_s = \phi f r \frac{f_2}{f_1}. \tag{26}$$

Substituting numbers in Equation (26) gives a value for $\delta_s = 5.7$ microns. This evaluation of resolution is somewhat optimistic, since it ignores other terms influencing the final size of the seeing disk, namely (a) intrinsic diffraction effects and (b) aberrations of the spectrograph lenses, as mentioned earlier. We will now estimates these two terms.

(a) The widening of the spectral lines due to diffraction is equal to

$$\delta_d = \lambda \frac{f_2}{d_2} \tag{27}$$

demonstrating that the final measure of sharpness of the lines in a spectrograph is only a function of the aperture ratio f_2/d_2 of the camera lens. At a wavelength of 5500 Å, we have already calculated $f_2 = 50$ mm and $d_2 = 25.4$ mm, thus

$$\delta_d = 0.55 \times 10^{-3} \times \frac{50}{25.4} \simeq 0.001 \text{ mm} = 1 \text{ micron}.$$

The blurring of the slit due to diffraction effects is much less than the blurring due to the seeing (the value of δ_s). Thus MERIS is not a *diffracted-limited* spectrograph. And in reality, not many spectrographs can claim to be so. (b) On the other hand, aberrations in optical components generally have a more significant impact on spectral resolving power. The term *aberration* covers most of the standard optical lenses defects: both spherical and chromatic aberrations, coma and astigmatism. Chromatic aberration is an important factor in spectroscopy that warrants careful consideration. It essentially means that the focus varies with the wavelength. In the best case, this means that the plane of the detector can be inclined in such a way that all wavelengths will be in focus. But in the worst case, the degree of chromatic aberration is so high as to require minor adjustments to achieve precise focus for the different wavelength ranges. With standard SLR photographic lenses this latter case is the most common.

Camera manufacturers obviously do not optimize their mass-market consumer photographic lenses for the UV and infrared regions. Thus to record a spectrum in the deep blue-violet and then the red, adjustments to obtain the best focus are essential. This implies that some simple and practical means of adjusting the focussing ring on the camera lens really needs to be built into the spectrograph design, either manually or (better still) by some motorised system.

For observations on real stars with the lenses considered here, we can estimate that the collimator will produce an image from a point source object, δ_c with a

FWHM of about 10 microns. For the camera lens we assume $\delta_o = 18$ microns. The difference comes from the fact that the camera lens is used with a larger f-stop than the collimator lens and so is more sensitive to optical aberrations.

Equation (19) can now be fully expressed as:

$$\delta_t = \sqrt{0.328^2 \times (0.0174^2 + 0.010^2) + 0.018^2 + 0.001^2} \simeq 0.019 \text{ mm} = 19 \text{ microns.}$$

The major contribution to spectral blurring comes from the camera lens (the term δ_o is 18 microns), even if we are using a good quality lens. In many spectrographs, line sharpness is often defined by the optical quality of the camera lens. If money has to be spent on improving spectral resolution, then it is on the camera lens that it should be invested.

Spectral sampling with 9 microns pixels meets the Nyquist criterion (two pixels at least to cover a resolution element), but there is not much margin of error $(2 . p_\lambda = 18$ microns). Equation (20) is therefore appropriate for calculating the resolving power at 5500 Å, and

$$R = \frac{0.55 \times 10^{-3} \times 1 \times 600 \times 50}{0.019 \times \cos 8.95°} \simeq 855$$

MERIS is a typical spectrograph in the class of medium resolution instruments, with a resolving power of around 7 Å in the green part of the spectrum. To obtain this result, divide the wavelength by the resolving power defined in Equation (2).

Now we have already seen the important influence that atmospheric seeing? Can have on the spectral resolution for a given spectrograph configuration and the necessity of analysing its effects. Let us now vary the seeing from 1.5 to 6 arcseconds in Equation (19). This typically covers the range from a steady to a highly turbulent atmosphere. In the high turbulence range, it is easy to verify with the equation above that the resolution for MERIS only changes by about 10%. This excellent result is due to the ratio between the focal length of the collimator and camera lenses. MERIS behaves like a powerful focal reducing lens, (of factor 3) and in so doing any image defects at the focal plane of telescope, when projected onto the detector, are reduced in the same proportion. More precisely, the reduction of the image spot between the focal plane of the telescope and the detector plane along the spectral axis is equal to $r \, f_2/f_1$. In this case, the anamorphic factor r works for us and not against us, its value being less than unity, according to the choice we made earlier for angles α and β.

Employing a slit width narrower than the seeing disk is sometimes needed, however, when we have a spectrum without any clearly identifiable, and therefore assignable, spectral lines that can be used for wavelength calibration. When this situation arises, we must use a spectral calibration lamp that produces a line emission spectrum, recorded before or after the target object spectrum. Both for the object and the calibration lamp, the light path through the spectrograph should be the same. This is why it is important to achieve precise focus of the target object at the focal plane of the telescope and positioned on a narrow slit. Such a slit is also essential when studying diffuse objects of finite size (*e.g.* nebulae, comets) or

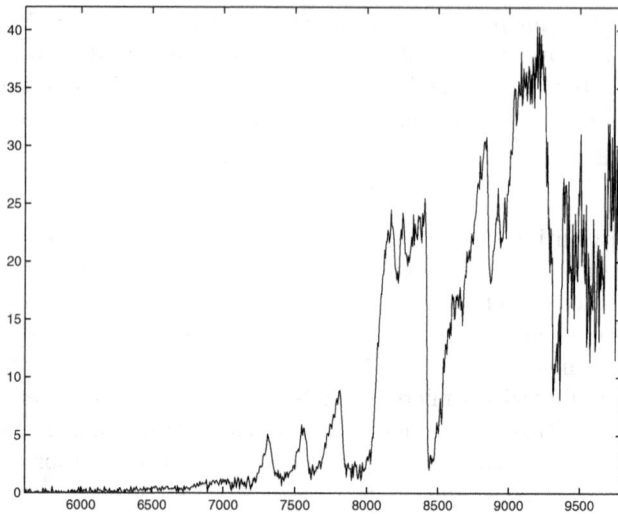

Fig. 4. Spectrum of the peculiar object V838 Mon, made with a 0.6 meter telescope at the Pic du Midi observatory, with a MERIS type spectrograph, recorded on 23th of April 2004. At the time this picture was taken the star was mainly emitting in the infrared. 9 stacked spectra of 2 min. exposure each. The CCD camera was an Audine KAF-0402M with a dispersion of 5.8 Å/pixel.

when the target object must be isolated from the neighbouring background (as in the case of a star embedded in an emitting gas cloud) to avoid overlapping spectra.

3.3 Limiting magnitude

In order to calculate the limiting magnitude for a given exposure time we must first determine the signal to noise ratio (S/N). The detection threshold for a spectrum is reached when the S/N > 5, *i.e.* five times greater than the noise total level. However, if we want to analyse spectral data with any degree of reliability and reproducibility, it is preferable to aim for S/N ratios of about 100 or higher.

In this section we will describe the theoretical basis that can predict the expected signal level, the noise and the signal-to-noise ratio (S/N) for a MERIS-type spectrograph under certain conditions.

The overall photometric efficiency (or *throughput*) of the telescope-spectrograph combination is the difference between the number of the photons actually emitted by the object under study for a given exposure time, computed from the top of the Earth's atmosphere and covering a surface equivalent to the telescope aperture, and the number of photoelectrons that fall onto a CCD pixel during the same time. This definition includes the effects due to atmospheric transmission. The throughput at a wavelength λ can then be written as

$$B\left(\lambda\right) = \left(1 - \varepsilon^2\right).T_a\left(\lambda\right).T_o\left(\lambda\right).T_s\left(\lambda\right).\eta\left(\lambda\right) \tag{28}$$

Table 1. Optical transmission of the telescope.

Optical element	Tansmission by surfaces	Number of surfaces	Total by element
Mirror	0.93	2	0.90
Corrector plate	0.98	2	0.96
		$T_0(5500)$	0.87

Table 2. Optical transmission of the spectrograph.

Optical element	Tansmission by surfaces	Number of surfaces	Total per element
Collimator lens	0.99	8	0.86
Objective lens	0.99	12	0.96
Camera window	0.96	2	0.92
CCD window	0.96	2	0.92
Grating	0.60	1	0.60
		$T_s(5500)$	0.42

where T_a is the atmospheric transmission, T_o, a factor characterising the optical transmission of the telescope, T_s, a factor characterising the optical transmission of the spectrograph, η is the quantum efficiency of the CCD detector and ε is a factor that corrects for the central obstruction of the telescope (this is zero for a refractor).

As an example, let us compute the throughput for a MERIS-type spectrograph around 5500 Å assuming the telescope is a Schmidt-Cassegrain (ε=0.33) with a main mirror of diameter $D = 200$ mm. The optical transmission typical of the various optical elements of the telescope is summarised in Table 1. In a similar way, the Table 2 shows the optical transmission performance of the spectrograph.

For an observatory situated at sea level and an object at an angular height of 45° above the horizon, the atmospheric transmission T_a is of the order of 0.63. The Kodak KAF-0402ME chip has a quantum efficiency of 0.68 at $\lambda = 5500$ Å. Applying Equation (28), the total throughput of the above setup becomes

$$B(5500) = (1 - 0.33^2) \times 0.63 \times 0.87 \times 0.42 \times 0.68 \simeq 0.14.$$

Although at first glance a result of 14% appears disconcertingly low, this value is, in fact, very reasonable for this class of instrument[8].

Now, at the detector the photon flux from the target object is spread not only along the spectrum (the line axis of the CCD image), but also along the perpendicular direction (the column axis of the CCD). We will explain the reason for

---

[8]For many years when spectrographs employed the photographic plate as the detector, much lower throughputs were the norm, of the order of 1%. Nowadays, the overall efficiency of the world's best professional instruments operating at high-altitude sites is still only about 30%.

this "vertical" spectrum lengthening and the advantage that it can give us, further below. For the moment, we must consider that the number of photoelectrons $N(\lambda)$ collected at one point of the spectral profile of the object is measured in the (horizontal) direction of dispersion and that by adding all the signal along the corresponding line of the CCD sensor, we therefore have

$$N(\lambda) = \frac{\pi}{4} E_0(\lambda) D^2 B(\lambda) \rho t\, 10^{-0,4\,V} \tag{29}$$

where $E_0(\lambda)$ is the spectral flux of a star magnitude $V = 0$ above the atmosphere in units of photons per square centimetre per second. For a star of spectral type A0V we have $E_0(5500) = 960\,\text{photons/cm}^2/\text{s}/\text{Å}$. If we then calculate the number of photoelectrons for a 300 second exposure time t for a star of this spectral type and a magnitude $V = 11$ (recall that we have already determined the inverse spectral dispersion ($\rho = 2.96\,\text{Å/pixel}$), the throughput result becomes

$$B(5500) = \frac{\pi}{4} \times 960 \times 20^2 \times 0.14 \times 2.96 \times 300 \times 10^{-0.4\times 11} \simeq 1500\, e^-.$$

If a narrow slit is used, the throughput is reduced by an amount that depends on the relative size of the seeing disk ϕ and the angular width of the slit ω. The slit transmission factor, T_f, must also be included in the throughput calculation for B. Assuming a Gaussian shape for the star image flux distribution, the slit transmission factor becomes

$$T_f = erf\left[\frac{\sqrt{\ln(2)}\ \omega}{\phi}\right] \simeq erf\left[\frac{0.8326\ \omega}{\phi}\right] \tag{30}$$

where erf is a special function called the *Error Function*, defined as

$$erf(x) = \frac{2}{\sqrt{\pi}} \int_0^x \exp\left(-h^2\right)\, dh \tag{31}$$

Table 3 below provides slit transmission factors for some typical values of the ratio ω/ϕ and the corresponding loss of light flux converted into a magnitude difference, Δm.

As soon as a narrow entrance slit is used with the spectrograph there is high risk of observing the throughput fall considerably just from simple geometrical considerations. Furthermore, the added difficulty of maintaining a star image on the slit during the whole exposure must not be forgotten. The throughput also depends on guiding quality. As a final point, if the slit is too narrow, the associated diffraction effects around the jaws of the slit also contribute to reducing the throughput. Diffraction will tend to increase the angle of rays exiting the slit as its width becomes narrower, and this will therefore increase the width of the beam. At slit width values lower than a certain limit, the light flux profile overflows beyond the diameter of the collimator lens and this light cannot reach the detector. For this reason, the slit width should not be smaller than the a minimum value, w_{min}, given by the equation

$$w_{min} = 2\,\lambda\,\frac{f_1}{d_1} = 2\,\lambda\,F_t. \tag{32}$$

Table 3. Transmission T_f of a slit and corresponding loss in magnitude Δ_m.

ω/ϕ	T_f	Δ_m
0.10	0.09	2.6
0.25	0.23	1.6
0.50	0.44	0.9
0.75	0.62	0.5
1.00	0.76	0.3
1.50	0.92	0.1
2.00	0.98	0.0

With a MERIS type spectrograph and a telescope working at f/10, the practical lower limit to slit width is $w_{min} \simeq 10$ microns.

The total noise associated with a single image pixel is the quadratic sum of the signal coming from the object, from the noise due to the sky background and from the thermal signal and the readout noise of the CCD amplifier electronics. An important consideration that is relevant to spectroscopy is that the recorded sky background signal is very often quite low for a fairly narrow slit width (but without being so narrow as to reach the equivalent width of the seeing disk). For example, a 100 micron slit is large enough to allow all the photon flux from a target star to pass through the spectrograph but at the same time is small enough to block a large quantity of the sky background signal surrounding the star. The sky background level recorded decreases in proportion to the slit width. Because of this, the photon noise of the sky background is often negligible in spectroscopy and therefore it is possible to perform spectroscopic observations under bright suburban skies and even under city lights. This is a distinct advantage when compared to traditional deep sky imaging techniques where the darkest possible skies are sought in order to maximise S/N.

The noise level, σ_t, in a pixel can be thus expressed as

$$\sigma_t = \sqrt{N(\lambda) + t\, n_y N_d + n_y n RON^2} \qquad (33)$$

where N_d is the thermal noise in electrons per second per pixel, t the total exposure time in seconds, n the number of individual exposures stacked to give the total exposure time t and the readout noise, RON, of the CCD sensor in number of electrons.

The coefficient n_y is the spectral width along the axis perpendicular to the spectral dispersion (the column axis). As each point of the spectral profile is the arithmetic sum of the useful signal contained in these ny pixels, each pixel noise will be added quadratically, which explains the presence of the term n_y in Equation (33). To minimize this noise, it is highly desirable to form the spectrum on the detector that has the smallest possible width δ_y in the column direction. This transverse linear width δ_y is readily deduced from Equation (27) by removing the anamorphic coefficient since, in the column direction, the grating behaves like a simple mirror. We therefore find,

$$\delta_y = \sqrt{\left(\frac{f_2}{f_1}\right)^2 (\phi^2\, f^2 + \delta_c^2) + \delta_o^2 + \delta_d^2.} \tag{34}$$

From the above expression, and rounding up to a whole number, we have $n_y = 2\,\delta_y/p_y$, with p_y being the pixel dimensions along the column axis. The factor 2 in the latter expression is added to ensure that the full signal, that obeys a Gaussian spatial intensity distribution, is included. If MERIS is coupled to a telescope of 200 mm at $f/10$ (focal length = 2,000 mm) and if the seeing is 3 arcseconds, we find that $n_y = 7$.

Let us now calculate the numerical value of the noise level assuming that the thermal signal from the CCD sensor, N_d, is about 0.1 electrons per second[9], the total exposure time is 300 seconds, split into three exposures of 100 seconds each ($n = 3$), and that that the readout noise from the amplifier in the CCD camera is equal to 18 electrons:

$$\sigma_t = \sqrt{1500 + 300 \times 7 \times 0.1 + 7 \times 3 \times 18} \simeq 92\ e^-.$$

The first observation to make is that the detector itself contributes the greatest noise. Consequently, a CCD camera with low readout noise is even more important in spectroscopy than in deep sky imaging. Any increase in the exposure time of the separate exposures before stacking will reduce the value of n and therefore lower the noise. It should be kept in mind, however, that it is always useful to statistically process multiple exposures of the same target (e.g. by median averaging and stacking) in order to eliminate cosmic ray strikes and any other artefacts that may be present on individual image frames. The value of n_y can be kept small if binning along the column axis is employed at the time of acquisition. Most CCD camera software is capable of binning in this way. It must be emphasised, however, that it is much more difficult to correct spectrum defects after the vertical pixels have been binned since this operation is irreversible.

The S/N mentioned earlier is simply the ratio of the useful signal calculated from Equation (28) divided by the noise calculated from Equation (33), thus

$$SNR = \frac{N}{\sigma_t}. \tag{35}$$

As was mentioned at the beginning of this section, the minimum useful S/N that is able to reproducibly differentiate a useful spectrum from the noise should be greater than 5. That being said, for serious scientific analysis, S/N values > 100 should really be aimed for. With the example described above, we found a $S/N = 1500/92 \simeq 16$. The stellar spectrum was easily detectable without ambiguity, but its contrast relative to the noise level is relatively low. In this case, the overall

[9]This is a typical value for the Kodak KAF-0401ME chip cooled to about $-10\,^\circ$C.

exposure time has to be increased considerably. To consider an example, it can be shown that a S/N of 100 on a magnitude 11 star can be achieved by stacking 12 exposures of 300 seconds each. This amounts to a total exposure time of about one hour.

S/N as we have just defined it applies to a spectrally sampled element, in other words to a "point" in the spectrum whose spectral width is equal to that covered by a single pixel. A different definition that is more appropriate for subsequent spectrum pipeline processing, and therefore of more interest to us here, consists in computing the S/N per element of spectral resolution. In this case,

$$SNR_{\Delta\lambda} = \sqrt{\frac{\Delta\lambda}{\rho}} \frac{N}{\sigma_t}. \tag{36}$$

The MERIS spectrograph samples the spectrum in steps ρ of about 3 Å by pixel, whereas the spectral resolution $\Delta\lambda$ is around 7 Å. Therefore the SNR per spectral resolution element for a magnitude 11 star with a 300 second exposure is given by,

$$SNR_{\Delta\lambda} = \sqrt{\frac{7}{3}} \frac{1630}{92} \simeq 27.$$

The calculation of S/N as we have described it in the previous paragraphs applies to a stellar continuum. However, many astronomical objects exhibit strong emission lines. These lines can exceed the continuum background by more than one order of magnitude (this is the case for various novae and some Be variable stars for example). Figure 2 shows one typical example where the detectivity of the Hα line is far better than the *detectivity* of the continuum. In some extreme cases, only the emission line will be visible, and the continuum will remain below the detection threshold.

Finally, we should note here the capability of operating a MERIS-type spectrograph in slitless mode (where $T_f = 1$). With MERIS in slitless mode we are able to record the spectra of very faint objects with smaller telescopes such as modest sized refractors. The two parameters, telescope and spectrograph, are intimately linked: with a telescope of small diameter, it is possible to omit the slit (because atmospheric seeing will have a minor impact on spectral resolution), and without a slit, objects of lower magnitude can be studied, with all the light rays collected by the telescope being passed through the spectrograph.

4 LORIS: A low resolution spectrograph

The LORIS spectrograph (for LOw Resolution Imager Spectrograph) is optimized for the studies of faint objects, such as novae, cataclysmic variable stars and comets. To reach such faint targets with small telescopes it is essential to work at relatively low resolution, in the range of 150 to 300. Transmission diffraction

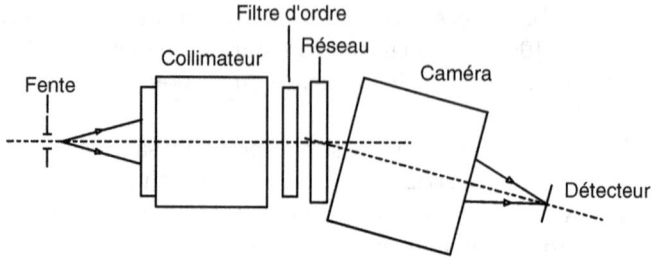

Fig. 5. Arrangement of the spectrograph elements based on a transmission grating. We notice the presence of an order filter just after the collimator. This red-orange filter of high-pass type eliminates the blue part of the order 2 spectrum when the order 1 infrared region has to be recorded. Without it, the two spectra will overlap and thus becomes unreadable.

gratings[10] are particularly well suited to this type of spectrograph, where the requirements are for high throughput in a compact instrument. The optical configuration resembles the classical Kirchoff-Bunsen spectrograph that originally employed a prism: an entrance slit, a collimator, a grating (with this time the diffracted rays passing through the grating), and a camera lens to form the final spectrum on the detector. A schematic diagram for such a spectrograph can be seen in Figure 5.

An important difference with a LORIS-type spectrograph, relative to higher resolution instruments, is that in addition to a complete first order spectrum being recorded, the zero-order image can also be seen without having to rotate the grating. This has two distinct advantages: firstly, by opening the slit, it is possible to identify the target object in the same image field, which can sometimes be a field rich in other objects because the target object magnitude is low, and secondly, we can simultaneously record both a full first-order spectrum of the object and an image of the object itself, as just mentioned. This is an effective and practical way to calibrate the spectrum.

With LORIS, light rays reach the grating at zero incidence angle ($\alpha = 0$). Thus a simplified expression for the diffraction angle β as a function of the wavelength λ is

$$\beta = \arcsin\left(k\, m\, \lambda\right). \tag{37}$$

The spectrograph parameters are adjusted in order to have both the zero order image ($k = 0$) and the first order spectrum ($\lambda = 1$ micron, $k = 1$) simultaneously imaged on the available surface area of the detector. The LORIS spectrograph employs a Kodak KAF-1600 CCD sensor with 1526×9 micron square pixels along the dispersion axis. By choosing a 300 lpm grating we can easily determine that the focal length of the camera lens should not exceed 35 mm in order to record both zero and first orders at the same time.

---

[10]With transmission gratings, the grating pattern is etched on a transparent substrate and light rays are transmitted through the grating, hence the name.

With MERIS, the effects of seeing and atmospheric turbulence on spectral resolution can be restricted by ensuring that the focal length of the collimator lens is greater than that of the camera lens. For the collimator lens, LORIS employs a photographic lens of 50 mm, at $f/1.8$, and a camera lens of 35 mm at $f/2$. In this way, LORIS can be used for point source objects with a large slit width without any impact on spectral resolution (Nb. This mode of observation is still termed "slitless", even though we are employing an open slit to limit the sky background noise level). The transmission grating in this spectrograph was purchased from Richardson Gratings (now part of Newport Corp. http://www.gratings.newport.com/home.asp) This particular grating is square with dimensions of $25 \times 25\,\mathrm{mm}^2$, a thickness of 6 mm and has 300 lines/mm, blazed at a wavelength of 5800 Å. The centre wavelength is $\lambda_0 = 5000$ Å. From Equation (37) above we can derive the fixed angle between the optical axis of the telescope and the camera lens. Substituting in the numerical values gives:

$$\beta = \arcsin\left(1 \times 300 \times 0.5 \times 10^{-3}\right) \simeq 8.63°.$$

From Equation (16) the reciprocal dispersion may be calculated and, in this case, $\rho = 8.5$ Å/pixel. From Equation (24), the resolution of LORIS for 3 arcsecond seeing conditions is $R = 300$.

Neglecting optical distortions the linear distance along the sensor itself, between the zero order image position and any point in the spectrum at a wavelength λ, is given by

$$\Delta x = f_2 \cdot \tan\beta. \tag{38}$$

For example, at a wavelength of 5000 Å

$$\Delta x = 35 \times \tan 8.63° \simeq 5.31 \text{ mm}.$$

With 9 microns pixels, the distance, expressed as the number of pixels, between the zero order image and the point in the spectrum corresponding to a wavelength of 5000 Å is equal to $5.31\,/\,9 \times 10^{-3} = 590$ pixels. This calculation can be generalised thus: if x_0 is the zero order position of the object along the dispersion axis, the wavelength for a given point at the coordinate x is

$$\lambda = \lambda_0 + (x - x_0)\,\rho. \tag{39}$$

The above equation assumes a constant linear dispersion model across the whole spectral range. This assumption is easy to challenge and is an approximation at best owing to unavoidable image distortions in the camera lens. Fortunately, it is still accurate enough for some key spectral features to be quickly identified. A more rigorous spectral calibration routine would require the use of a calibration lamp and a narrow slit.

Figure 7 shows the LORIS spectrograph mounted at the focal plane of a 60 cm Newtonian reflector at the Pic du Midi observatory in the Pyrenees in southern France. The overall throughput of this configuration is high, close to 20%, and the limiting magnitude is 16.8 in the red for a S/N = 10 with a one hour exposure.

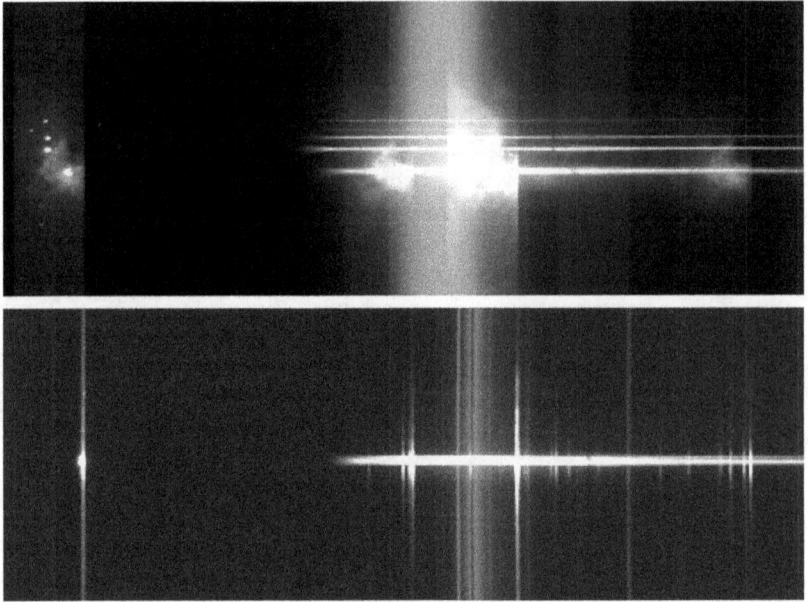

Fig. 6. The upper photograph image is a LORIS spectrum of the Orion nebula (M 42) with a widely opened slit. On the left hand side of this image we can see the white light image of the nebula (zero order). On the right is the corresponding spectrum. All the objects inside the slit produce a spectrum; this is the case, for example, for some of the stars which are visible in the field of view. In the lower image, the slit has been reduced to isolate only a narrow portion of the nebula. Classical lines of nebula are visible, such Hα, [OIII], etc. This spectrum has been acquired in an urban site. Emission lines visible all along the image come from the intense sky background.

5   LHIRES: A high resolution spectrograph

The best way to secure results that are of serious interest to the scientific community is to work with a high resolution spectrograph. This general opinion is based on outcomes from several workshops and conferences whose goals are to promote professional and amateur collaboration amongst astronomers. The interested reader can consult the various presentations made during the 8th Astrophysical Olron Summer School dedicated to PRO-AM collaboration in spectroscopy organised by the CNRS[11]. For high resolution studies, spectrograph resolving powers equal to or greater than 10,000 are very often needed. This requirement is achieved with the LHIRES spectrograph (short for Littrow HIgh REsolution Spectrograph) that is the subject of this section.

The technical characteristics of LHIRES have been refined and optimised so that spectra of magnitude 6 stars are recordable with a S/N = 80 after a one

---

[11]http ://www.astrosurf.com/aude/oleron/

Fig. 7. LORIS spectrograph at the focal plane of the Newton telescope of 60 cm at the Pic du Midi observatory.

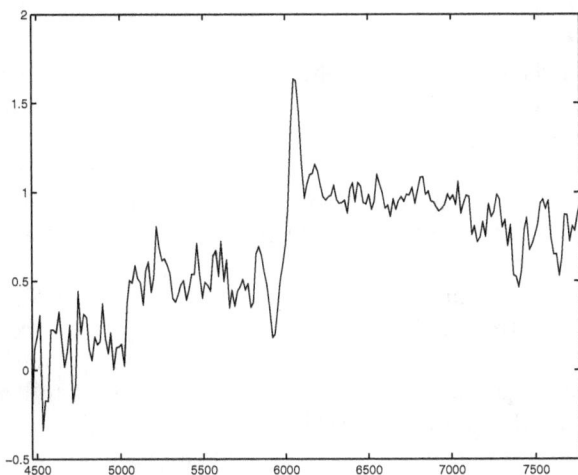

Fig. 8. A spectrum of Quasar APM 08279 + 5255 ($z = 3.87$) obtained in April 2003 with the LORIS spectrograph at the focal plane of the Newton telescope of 60 cm at the Pic du Midi. The exposure time of the magnitude 14.8 object is 38 minutes. The emission line visible at the centre is Lyman α, which is normally at a wavelength of 1216 Å (emission wavelength). This line, however, is actually observed at about 6000 Å due to the extremely high red-shift produced by this object.

hour exposure on an 8 inch (200 mm) telescope. The number of stars that fall within these design requirements total around 15,000 which is more than adequate for many serious investigations. Moreover, the type of telescope that has been targeted for this spectrograph design is the 8-inch f/10 SCT and similar types, that are widespread among amateur users. LHIRES is a spectrograph designed

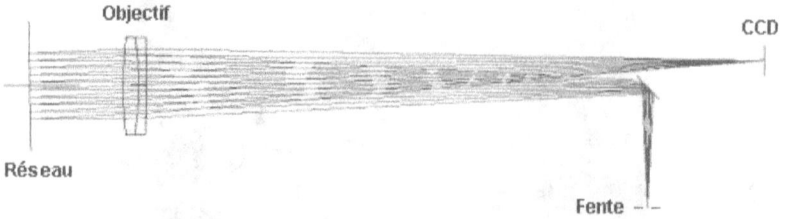

Fig. 9. Optical schematic of the LHIRES spectrograph. After passing through the slit, the light beam is deflected 90°, via a small plane mirror, toward the sole lens of the instrument. The slit is positioned at the focal plane of this lens, so that an image of the slit will be produced at infinity (*i.e.* parallel). Upon reaching the grating, the parallel beam is diffracted back again, passing through the same lens, and a spectrum is produced at the focal plane of the detector.

in the *Littrow* configuration, where the same optical element (in this case a lens) is used both as the collimator lens and the focusing lens (refer to the optical schematic 9).

A beam of light from the slit will pass through this lens twice. The first pass collimates diverging light rays from the slit into a parallel beam before they reach the grating, and the second pass then focuses a dispersed spectrum onto the CCD. In Littrow mode, the angles of incidence and diffraction (α and β) are almost identical. In practice, a slight difference is in fact needed to avoid the two beam intersecting exactly and overlapping, but this difference is small, of the order of around 2 degrees.

The single lens is a simple achromatic cemented doublet of 200 mm focal length with a diameter of 30 mm [If the telescope is working at $f/10$, we could, in principle, get away with using a 20 mm diameter lens, but this is a lower limit – see the Eq. (6)]. A small elliptical mirror is positioned after the slit to bend the beam by 90°. The mechanical axis of the spectrograph is then folded at right angles with respect to the optical axis of the telescope and can therefore easily pass through the tines of a fork mounted telescope.

The optical assembly is completed by a semi-reflecting glass plate that acts as a beam-splitter and is inserted in front of the slit. Its function is to direct about 10% of the light flux coming from the telescope toward a small camera to be used for guiding during an exposure. This guide camera can be a webcam or a video camera; currently a Watec 902H video camera is being used. The beam-splitter also allows light from a spectral calibration lamp to pass into the spectrograph. Calibration lamps produce an emission line spectrum with many lines whose wavelengths are known precisely. An excellent and inexpensive choice of calibration light source for the amateur are the small neon lamps that can be found in the lighting sections of supermarkets or hardware stores. Neon lamps in particular are suitable for calibrating spectra around the H-alpha line, for example.

The main problem with using a beam-splitter is that 10% of the light flux is effectively lost and not detected by the CCD camera. An alternative approach

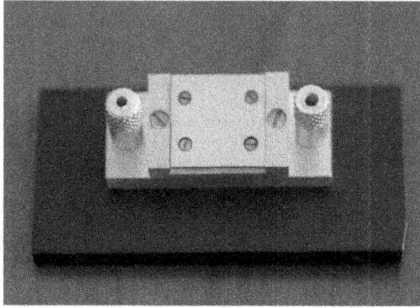

Fig. 10. Example of a reflective slit. The slit plane is tilted at an angle of 10° in order to direct the reflected beam towards the guide camera. The slit is made of aluminium, which is relatively easy to polish and exhibits a high degree of reflectance.

that is often employed with this type of instrument is to use a slit having highly polished jaws that reflect light that does not pass through. With the aid of a small relay lens or mirror this light is then focused onto the guide camera (see the Fig. 10). With this guiding mode the image of the star whose spectrum is being recorded is never actually seen. The goal here is to ensure that the star is never visible so that most or all of the photons go through the slit and contribute to forming the spectrum. A beam-splitter obviously unnecessary in this case. Spectral throughput in this mode is consequently improved somewhat, but a disadvantage is the additional requirement to fabricate and polish the reflective slit which can be complex process.

The main optical component of the LHIRES spectrograph still remains the diffraction grating. The model selected for LHIRES is a square holographic grating of 50 mm on a side, with $m = 2400$ lines/mm (from *Edmund Optics*). The beam reaches the grating at an angle of incidence of about 52° ($\alpha = \beta = 52°$). The two jaws of the slit are made from aluminium with precisely filed edges to make them as smooth and parallel as possible. The slit width w is set to 28 microns. The width is manually adjusted to this dimension and then fixed with locking screws. The projected slit width on the sky is 2.9 arc seconds if the focal length of the telescope is 2 meters. As opposed to the MERIS and LORIS spectrographs described above, the unit degree of magnification obtained with LHIRES ($f_1 = f_2$, $r \simeq 1$) means that a narrow slit is absolutely essential in order to prevent the spectral resolution becoming too dependant on atmospheric seeing conditions. In most cases, the slit is narrower than the diffusion spot due to atmospheric turbulence. Under such conditions, the resolving power is termed slit-limited. The projected width slit at the detector is then given by

$$\delta_w = w\,r\,\frac{f_2}{f_1}. \tag{40}$$

If we write $\delta_t = \delta_w$ in Equation (20), the resolving power in the slit-limited case then becomes

$$R_{slit} = \frac{\lambda\,k\,m\,f_1}{w\,\cos\alpha}. \tag{41}$$

Since the angular width of the slit on the sky is $\omega = w/f$, we obtain Equation (4) again and so produce the general expression for the resolving power of a spectrograph. Filling in the numbers for a wavelength of the Hα line ($\lambda = 6563\,$Å), we obtain

$$R_{slit} = \frac{0,6563 \times 10^{-3} \times 1 \times 600 \times 200}{0,028 \times \cos 52°} \simeq 18\,000.$$

In practice, the effective measured line width for a stellar spectrum with LHIRES gives a resolution of around 17,000, which is very close to the above theoretical value, the small difference being due to optical aberrations originating from the achromatic doublet lens.

The spectral range covered by a CCD sensor from the Kodak KAF-400 series of chips is about 90 Å with a sampling factor of 0.115 Å/pixel. Thus with a single exposure, only a relatively small portion of the spectrum is recorded. This limited free spectral range can be contrasted with the modern instruments now in use by professional astronomers where with a single exposure essentially the whole spectrum from the blue end to the red end can be recorded with the same degree of detail and resolving power as LHIRES. This latter technique is the field of Echelle spectroscopy and the spectrographs employed are known as Echelle spectrographs. Unfortunately, for the moment at least, the additional cost associated with designing an Echelle spectrograph is prohibitive for the majority of amateurs. In summary, with LHIRES only one or a small group of closely spaced spectral lines can be examined, but in great detail. It is as if we had used a magnifying glass on the spectrum.

The photometric throughput with LHIRES is about 5% if we include the atmosphere, the telescope, the entrance slit, the detector and assume seeing conditions of 3 arcseconds. Manufactured from sheet aluminium and other basic stock, LHIRES is relatively light in weight, weighing in at around 1.7 kg without the CDD camera. When we add a typical CCD detector to LHIRES, the weight increases to about 2 to 2.5 kg. This is an excellent result for a spectrograph that is capable of a resolution greater than 15,000.

Operating the LHIRES spectrograph is quite straightforward. The target star, which should be of sufficient brightness (and therefore easy to identify) is focused and centred on the jaws of the slit with the aid of the video camera. The star is then maintained in this position, again with the help of the video camera. The video signal from the guide camera is sent to a small monitor screen or, with appropriate software, directly shown on the screen of a laptop computer to track the star's position during the exposure. An auto-guiding system can also be envisaged and designed.

Unlike conventional CCD imaging where any guiding errors during exposure can potentially ruin a fine astrophoto, guiding errors made during the exposure with a spectrograph merely lead to a momentary loss in signal, since the light is temporarily not entering the slit. The signal is recovered once the target has been re-centred on the slit. The slit axis is most often orientated parallel to the R.A. axis of the mounting in order to minimise any periodic error in the telescope drive,

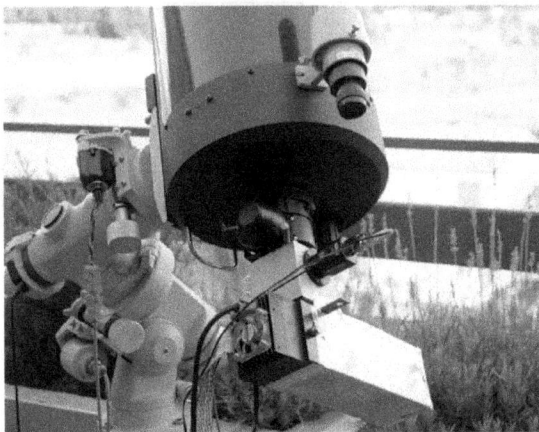

Fig. 11. LHIRES spectrograph at the focal plane of a Celestron 11 telescope. We can see the guiding video camera at the interface between the telescope and the spectrograph.

although the LHIRES spectrograph is quite tolerant of mount defects. Nevertheless, the better the guiding, the better the photometric throughput will be, and the greater the S/N for a given exposure time.

Figure 12 is typical of the results that can be obtained with a LHIRES spectrograph. The investigation in this particular example concerned the detection of lithium in the star's atmosphere. Lithium is well known to be one of the basic elements that formed during the first moments of the Big Bang and whose properties are a good indicator of the processes occurring during stellar evolution. This somewhat difficult observation has until now been the sole reserve of the professional astronomer, but with spectrographs such as LHIRES this type of result is now accessible for the serious amateur to attempt. The observation is a satisfying accomplishment in its own right, relating as it does the telltale spectral signature for this fundamental element and its astrophysical importance. But beyond this, other studies could be considered, such as monitoring the abundance of elemental lithium for any time-dependent changes in some objects.

Figure 13 shows spectra of the star Zeta Orionis recorded at different times over a 24-hour period. This star is well known since it shines close to the famous Horse Head nebula in Orion. The Hα line observations reveal that this particular supergiant is undergoing some very significant changes in its atmosphere.

Line profile distortions such as this are indicative of erratic ejections of matter, similar to the solar prominences observed on our own Sun, but on a much larger scale. With the spectrograph, this well known and easily observed object just becomes one more single point in the sky.

Be stars type are also a very interesting family[12] and should be of interest to the amateur spectroscopist. These objects are non-supergiant stars of spectral

---

[12]Refer to the relevant chapter by C. Neiner.

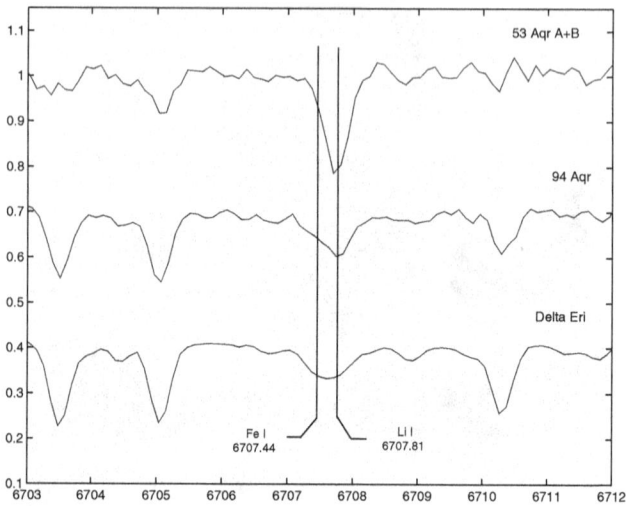

Fig. 12. The spectral signature of elemental lithium in three different stars. The line is at a wavelength of 6707.8 Å in the far red. The close proximity of the Li line to one due to Fe at 6707.4 Å indicates the importance of having a high degree of dispersion to ensure correct line identification. In these example spectra the separation between the Fe line and the Li line is not large enough to clearly measure the separation; the two lines are blended, but do nevertheless allow an estimate of line width that is linked to the abundance of this element. For example, in the spectrum of δ Eri the two lines have the same intensity. The observed wide, almost symmetrical profile, is the sum of the two signals. A simple mathematical deconvolution of the composite line profile allows us to calculate the lithium line width alone. Agreement with the results for 94 Aqr and δ Eri obtained at professional observatories is excellent. On the other hand, a significant difference is observed for 53 Aqr. These data were acquired using LHIRES at the focal plane of a 280 mm diameter Schmidt-Cassegrain catadioptric telescope.

class A or B in an evolutionary phase that exhibits, among other things, hydrogen emission lines, as opposed to the more usual absorption lines. The emission lines come from a disk of circumstellar material in rapid rotation around the central star. The origin of this disk, the mechanisms that produce and maintain it, and how the disk changes with time are all aspects of Be stars not yet fully understood and are the subject of considerable scientific research. Professional investigations on Be stars require more observational data in order to build appropriate computer simulation models for disk formation and evolution. Many of these objects are observable with telescopes of 200 mm to 400 mm diameter. Even the discovery of new Be stars could be considered through a survey program of normal B-stars for any changes in absorption behaviour. Figure 14 shows some typical spectra of Be stars around the Hα line.

A spectrograph such as LHIRES can even be targeted on our own star, the Sun, to perform time-based spectroheliography. With the spectrograph slit aimed

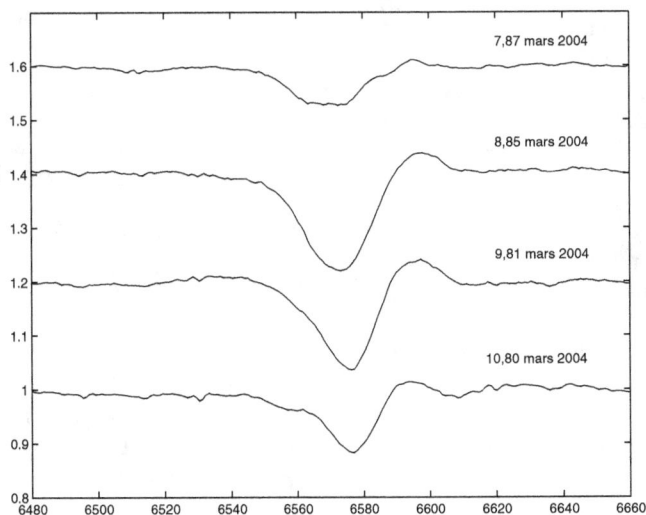

Fig. 13. Hα line evolution of Zeta Ori star within an interval time of 24 hours in March 2004. During the intense eruptive phase, already captured by the LHIRES, it is possible to record changes in from hour to hour.

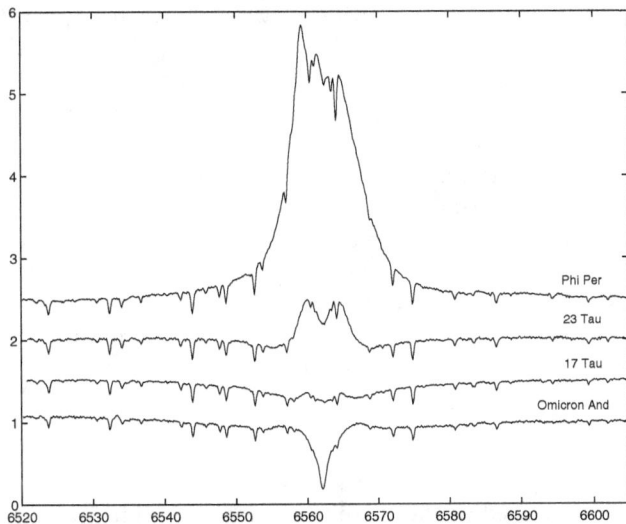

Fig. 14. From top to bottom, spectra of the Be stars Phi Per, 23 Tau (Alcyone), 17 Tau and Omicron And. The Hα line profile exhibits various shapes that can change over time periods ranging from a few hours to months or even years depending of the object. The much less intense and narrow absorption lines are due to atmospheric water vapour. All spectra were acquired under suburban skies of the city of Toulouse with LHIRES at the focal plane of a 0.28 m Celestron telescope.

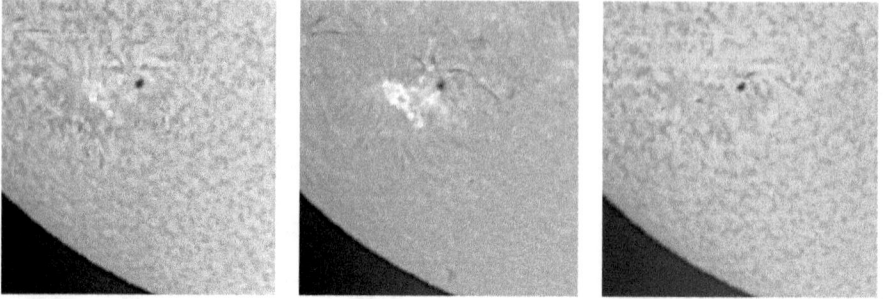

Fig. 15. Spectroheliogram of a part of the solar surface, obtained with LHIRES at the focal plane of a 128 mm refractor. From left to right, the Sun's image produced at a wavelength corresponding to the blue edge of the Hα line (at −0, 42 Å from the central peak), at the central peak of the Hα line and on the red edge of the Hα line (at +0.42 Å from centre). The camera is a mere webcam, type Vesta Pro. The spectrum is sampled at 0,072 Ångströms/pixel.

to one side of the solar disk, if we stop the sidereal motion of the telescope, the Sun's image will slowly pass through the slit position as the Sun drifts across the telescope field. At regular time intervals the spectrum is recorded in order to produce an image of the Sun in monochromatic light. In this application LHIRES is operating as a very compact spectroheliograph.

Data acquisition rates for this application need to be quite high. A video camera or webcam is used and the video signal downloaded in real time to a personal computer using a high speed USB cable (see Fig. 15) and a monochromatic image of the Sun built up afterwards. Assume for example that n spectra were acquired during the passage of the solar image over the entrance slit. The next step is to measure the signal at the centre of a particular spectral line. The information, collected for all n spectra, is an instantaneous monochromatic slice of that part of the solar disk along the slit. This represents the monochromatic image of one column in the final image. The completed image will thus have n columns and the height of the full image in pixels is equal to the slit height[13].

6 Conclusion

This chapter has described a range of different spectrographs that can be constructed and used by amateurs and has gone into some details to describe the design rules. In their respective fields of application, each spectrograph is capable of generating scientifically valuable data. The simplicity of approach does not limit the utility and interest of these instruments. The design approaches described

---

[13] Usually, some additional geometrical correction factor is necessary in order to ensure the Sun's image is perfectly round.

here underpinned with the basic theoretical principles can be copied, eventually improved upon, and all at low cost. This strategy opens up whole new horizons for professional-amateur collaboration and coordinated programmes of research. Moreover, the possibility of acquiring scientifically usable spectra even under light-polluted urban skies must again be emphasised. This has to be an important factor of interest to the serious amateur wishing to enter this fascinating field.

The discipline of astronomical spectroscopy has only recently become of real interest to amateurs. This is somewhat surprising since, with the ready availability of highly sensitive CCD detectors and the right optical components, the technical obstacles to success in spectroscopy are no longer an issue. The future should certainly see the field be better developed within the amateur community and this is already beginning to be demonstrated around the World as word spreads. As an example the work done by the ARAS group can be followed (Astronomical Ring for Access to Spectroscopy[14]).

---

[14]http://www.astrosurf.com/aras/

Astronomical Spectrography for Amateurs
J.-P. Rozelot and C. Neiner (eds)
EAS Publications Series, **47** (2011) 73–101

SPECTRAL DATA REDUCTION

V. Desnoux[1]

Abstract. Spectral data reduction is a crucial step, as so important as the acquisition process. Rigor shall be as well applied at this stage to not degrade the quality of the data acquired. It is the last step before the computation of the astrophysical quantities. It is thus necessary to understand what is involved in each processing steps and the underlying limitations. Thus limitations or limit conditions are usually embedded and then hidden in a specific software command that everyone use without completely mastering what is really at stake. In this chapter we will introduced the different type of processing and spectral analysis which can dispose an amateur in several software packages. The first processing will treat about basic images correction which are linked to the CCD acquisition and the processes pipeline. We will then describe the different strategy to calibrate the spectral profile in wavelength and the few corrections driven by the earth environment to increase the accuracy of the future measurements. We will end with the description of few tools on quantitative spectral analysis and their limitations. The practice of spectral acquisition and processing is just at its beginning in amateur world. We hope that this section will demystify the essential step of spectral processing and will motivate the amateur to go beyond the contemplative watching of sky wonders.

1 Data reduction, fundamental steps

This section describes the mandatory steps of image processing before any further spectral processing.

- Thermal signal and offset corrections

- Flat-field, and parasitized pixed corrections

- Geometric distortion correction

---

[1] Association Aude
e-mail: `valerie.desnoux@free.fr`; `http://valerie.desnoux.free.fr`

© EAS, EDP Sciences 2011
DOI: 10.1051/eas/1147003

- "Sky background" correction

- Spectral profile reduction

- Wavelength calibration.

1.1 Pre-processing of a spectral 2D image

Offset and dark. The spectral image is not really different from a classical image for all the sensor-driven pre-processings. First corrections shall subtract the offset and the thermal signal on raw image. Offset signal, so-called "bias" signal is the level of the pixels when reading a "zero-length" exposure. The offset image is withdrawn from the raw image. The offset image is obtained as for example by the median sum of a series of n images at zero second exposure time to reduce the noise. The thermal signal is signal generated by the sensor itself and proportional to the time and the temperature. The thermal image, so-called "dark" image is obtained as for example by the median sum of series of n images of "black", *i.e.* no signal reaching the sensor, at the exposure length equal or close to the exposure length of the raw images. This correction withdraw the level of the thermal signal but does not remove the noise. That's why acquiring a series of image and doing the median will decrease the noise. The dark image shall have its offset corrected then to be the true "proportional to time and temperature" signal component. A "hot-pixel" or a "dead-pixel" is a pixel for which the response is not proportional to the signal received. It can be detected as it exhibits a value which is clearly out of the range of the average of the pixel values in a dark image. The detection algorithm will build the pixel values histogram and the user will declare a threshold beyond whose the pixel will be declared as "dead". To correct such pixel, a classical algorithm replace the dead pixel value by the average of its neighbors. Some other effect like cosmic rays can also alters a pixel values, which also become "out-of-range", the same detection and correction algorithm will also apply for this type of artifacts. Not correcting them, would result in a very fine but false spectral line.

Flat-field. In standard imaging, a flat-field image is used to compensate the local non-uniformity of the image created by defects in the optical and sensor image chain. (dust, local non-uniformity of the pixel response). To get the flat-field image, a uniform field image is taken. Due to the defects, this image appears to have local non-uniformities, a map of the defects, of the local gain non-uniformity. By dividing the image by its flat-field gain map, the local attenuation are locally "boosted" and thus compensated. In spectroscopy, the concept of uniform field is a little more complex as the sensor sees behind the spectrograph the spectrum of the uniform field. But as for standard imaging, local attenuation or artifacts has to be be corrected to avoid false lines or continuum alteration generation. A uniform field may not have a "uniform" spectrum, it depends of the light source which has produced it. If the flat-field are taken from moderately luminous sky the recorded spectrum would show the solar and the earth atmosphere spectrum.

Fig. 1. Flat-field acquired by the illumination of a white paperboard by an incandescent lamp. The non-uniformity of the image is the combined effect of the spectral response of the sensor and the local attenuation or pixel non-uniformities.

Far from being uniform. So, flat-fields in spectroscopy are usually produced by lightning a white uniform surface by a lamp which produce an "as continuous as possible" spectrum, meaning exciting no absorption or emission lines. At this point, one would be tempted to simply take this image as the flat-field. There is one additional effect in the sensor which has to be considered. The response of a pixel will also depend of the wavelength of the light which reach it. Sensors can be more sensitive in red range of the spectrum and much less in the blue part. This effect has to be separated here to strictly eliminate the local attenuation of the image chain and the local non-uniformities of the pixel gain. The spectral response correction of the sensor will be addressed separately in another specific spectral processing further on.

To get to the local non-uniformity gain map, first step is as described above to take an exposure of a uniform field lit by a lamp whose the spectrum has no absorption or emission lines. Incandescence lamps are good candidates. Among them, tungsten filament lamps produce a "no-line" spectrum which will be more intense in the red part of the spectrum than in its blue part.

If we consider the intensity profile of a line, along the x-axis of the image, the intensities are higher in the red part of the spectrum. It is the combined effect of the lamp spectrum which is more intense in this range, and the good response of the sensor also to red light. But are also visible on this intensity profile, local variations which are produced by pixel to pixel gain non-uniformities or local dust attenuation. Those are the local gain variation which has to be captured in the flat-field gain map.

To focus on the local gain variation, the slope or the slow variation of the intensity shall be removed. To remove it, this effect shall be first isolated (smoothing function).

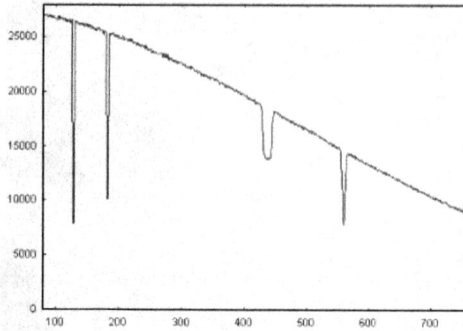

Fig. 2. Intensity profile of an image line, spectral region around H-alpha. Artifacts due to dust or other sources are visible – (simulation).

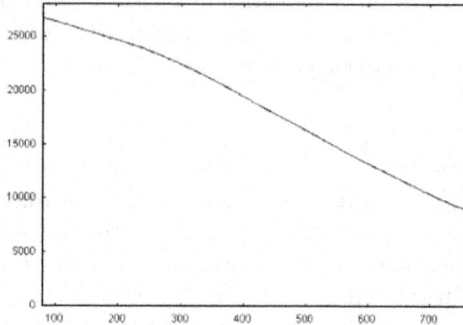

Fig. 3. Resulting profile after smoothing: overall gradient of the continuum.

By dividing the original profile by this smoothed gradient profile, only the local and rapid gain variation will be retained. This is the local "gain map" which we were looking for.

What we have shown with one line of the image, software does these processing on the whole image. The resulting image will be corrected for thermal signal and offset and be kept as the "flat-field" image.

In the pipeline of the pre-processing of a spectral image, the raw image will first have its offset map subtracted, then its dark image and be dived by the flat-field.

1.2   Registration

For weak objects, the exposure time required to get a correct signal to noise ratio are usually quite long. And specially longer in spectroscopy where light is spread across the sensor. Long exposure time in a one-shot image are not recommended with amateur conditions. It is much better to break it into a series of shorter exposure and later on added. It is less painful to eliminate a one-minute image which exhibits guiding errors in a ten images sequence, than throwing away a ten

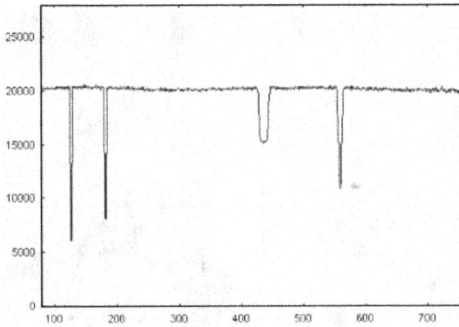

Fig. 4. Raw profile divided by the smoothed profile.

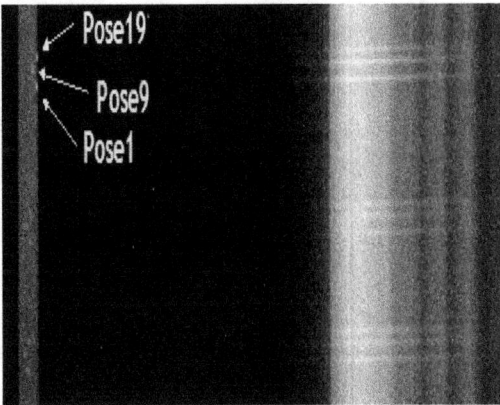

Fig. 5.

minutes exposure image... This way of acquiring faint object is very standard in classical imaging for amateurs. Each images are then summed up to recover the signal to noise ratio. Before the summation, each image are carefully registered, if not, a small defect in the guiding will make the star appear as a dotted line on the resulting image or blurred, which will degrade the final resolution.

In spectroscopy, the process is the same, but to register the images, there is no star, the spectral image usually does not include the zero-order which is the image of the star (eventually through the slit if there is one). The registration algorithm will have to register on the spectral lines themselves, in absorption or in emission or will require that the user will manually shift images by observing the image of the differences of two successive images on the whole series. Another option for this processing is to differ it and process each raw image of the sequence, reduce them up to the 1D profile (binning step) and register the 1D profiles before the final addition.

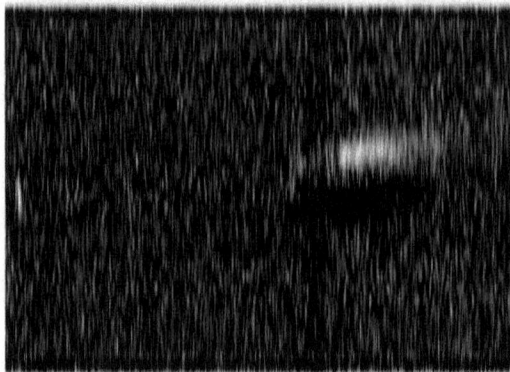

Fig. 6. Image of the difference of two successive images, without shifting.

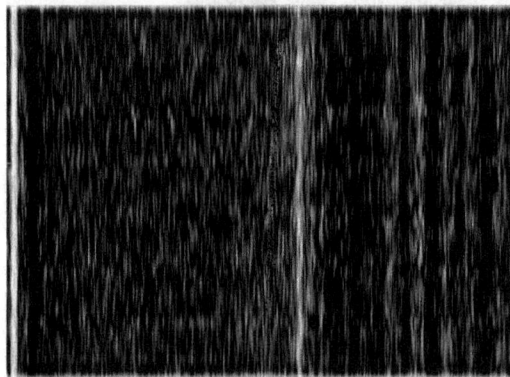

Fig. 7. Image of the difference, after shifting – the object spectrum disappear, the slit image appears in relief.

1.3   Geometric corrections

If attention is paid to the shape of the lines in the raw spectrum, it is not unusual to see these lines not strictly straight and aligned with the Y-axis of the sensor. The shape can be altered by the spectrograph optical configuration and its optical limitations. If those distortions are not compensated, when reducing the 2D image into a 1D profile, the lines will be blurred, and the resolution will be degraded. The conversion into 1D profile usually involve a column by column intensity addition across the spectral image. If lines are tilted or spread across several columns, this will degrade the final resolution.

This blurring effect will also affect the continuum of the spectrum, mixing pixel values of different wavelength.

Multiple geometrical distortions may be observed in real condition. As they come from the specific configuration of the spectrograph, they have to be defined

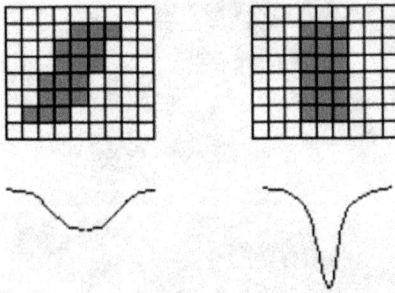

Fig. 8. Line inclination effect on the resolution after profile reduction. *Left*: inclined line and its resulting profile. *Right*: a perfectly aligned line, and its resulting profile.

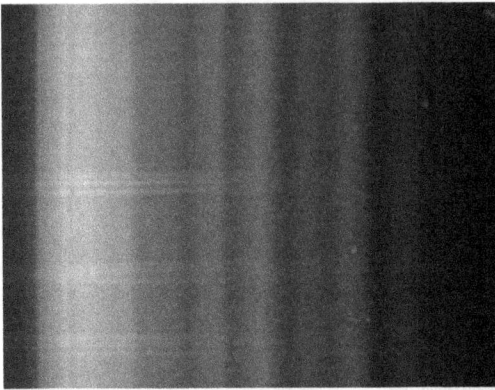

Fig. 9. Image with "smile" distortion.

and corrected for each type of configuration. In spectral processing software, multiple tool to first evaluate the defect then fix it are available.

Here are some classical effects:

"Smile" effect. Spectral lines does not appears straight, but bowed. The correction will apply a shift, line by line. The shifting value are computed from a spherical model of the smile effect for which the center and the radius will be provided.

"Tilt" effect. The slope of the continuum along the horizontal axis is corrected by a rotation of the image. The angle of the rotation has to be provided.

"Slant" effect. This effect affects the lines but not the overall spectrum continuum. To rectify the lines, and not the overall spectrum, a shift line by line is applied. The shifting quantity is computed from an inclination model for which the angle will be provided.

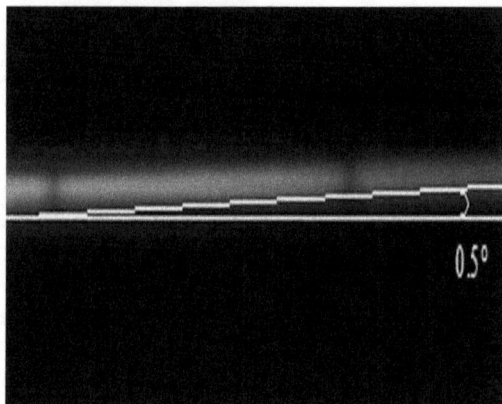

Fig. 10. Image with a spectrum having a tilt effect.

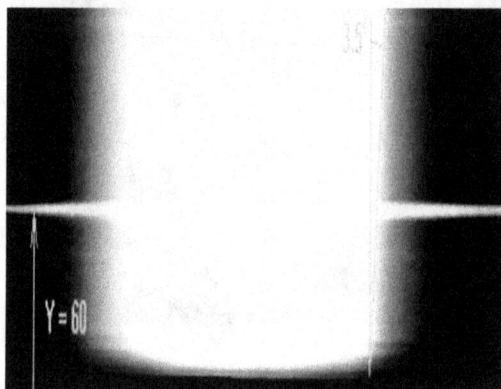

Fig. 11. Slant effect: image before correction.

Values of correction are very often determined by a trial and error process. But once determined, the correction shall be applied once on the raw image. A rotation of 2 degrees shall not be done with two steps of one degree rotation. The successive interpolation processes would alter the image definition.

1.4 Sky background removal

The informations contained in the spectral image is not limited to the spectrum of the object. The light from the entire aimed sky region will also be decomposed by the spectrograph. This means the sky spectrum superpose to the object spectrum. The sky spectrum can be or less intense, depending on the observation site and atmospheric conditions. In urban site, the sky spectrum can include the spectrum of the urban lamps around. The moon presence, reflecting itself in a local fog or high altitude clouds will add a spectrum of solar type to the background.

Fig. 12. Slant effect: image after correction.

Fig. 13. Raw spectrum of the object with visible sky background spectrum around.

Several correction methods can be used to compute the profile of the sky background spectrum. The most simple is to take the average value of the pixel intensities on a given area from each part of the object spectrum, where there is no object signal. This average value by column will then be withdrawn to each of the image pixels, column by column. A more sophisticated way is to compute not the average but the median to better filter possible artifacts. And even more, a gradient can be computed from the two region from the top and bottom of the object spectrum: this linear profile will then be removed from the raw spectrum object.

What is important is not to skip this step. Sky background lines can be very intense as said. During high altitude observations, at Pic du Midi observatory (2850 m), the emission line of fluorescence of the oxygen is quite strong and superpose to the faint spectrum of the objects. Only the sky background removal process will eliminate it.

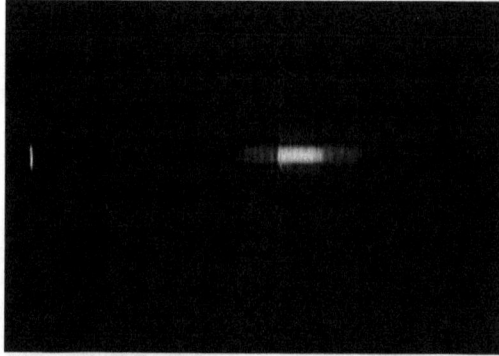

Fig. 14. Object spectrum after the removal of the sky background spectrum.

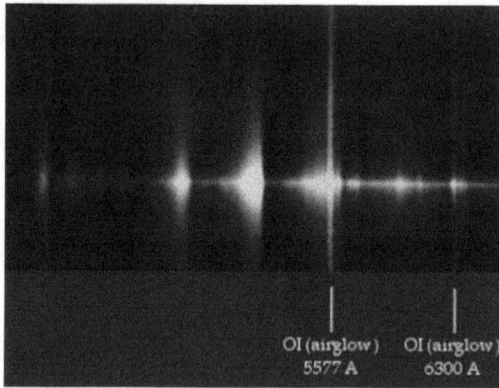

Fig. 15. Comet C/2001 A2 (Linear) spectrum before the sky background removal – the fluorescence line of atmospheric OI is quite visible.

1.5 Spectral profile reduction

At this stage of the processing, interfering signals coming from optical configuration, sensor and light pollution are wiped off. Only the object spectrum is kept, but spread across multiple lines of the sensor. The next step consists in converting this 2D-image into a 1D-profile. Why this? A star spectrum is by definition the graph of the intensity in each color, or more scientifically said, each wavelength: it is a one dimensional profile. The dispersion across the Y-axis is only due to the imperfection of the acquisition chain: the star image is not a point, the guiding is not perfect, the atmosphere and the turbulence blur the light beam. In the hypothesis of a non-unidimensional object, like a planet, the sun, the moon, this reduction process shall not be applied, it would blur the spectrum of each portion of the surface imaged by the slit. But in this case, those objects are luminous enough that each line has by itself a good signal to noise ration. For a faint star,

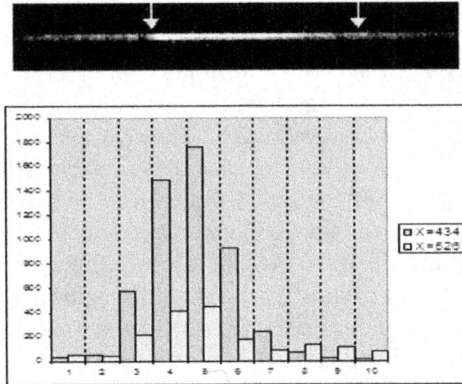

Fig. 16. Transverse section representing the pixel intensities for two different columns.

each line will contain the same informations than the other lines, but the signal is so low that considering the signal of only one line will give a very noisy profile. The "trick" is to combined the informations of all the spectrum lines by simply adding column by column the intensity across the Y-axis. This process is so-called "binning". To do this, the first question is "how to define the bounds of the spectrum"? Consider the section of the spectrum along the Y-axis, the transition between spectrum signal and the background is not conclusive. Adding a line from the background will add more noise than signal to the final profile.

- A manual selection of the binning area is an acceptable solution but not an optimal one. The process is to narrow the thresholds enough to visually estimate the transition position. Thus the software will use the top and bottom position to sum up the intensities by column.

- A second method, developed by the author let the software determine the optimum transition positions. The criteria used is the signal to noise ratio. At a first step, the algorithm compute the average spread out profile, along the Y-axis. Then, it order the lines by decreasing mean intensity. Finally, for each line, it will compute a signal to noise criteria and conclude if this line shall be added or not to the others.

Rejection criteria (Automatic binning in *Visual Spec* software from the author):

$$\Sigma L_n / \sqrt{(n)} < \Sigma L_{n+1} / \sqrt{L_{n+1}}. \tag{1}$$

Other categories of method, largely used by the professionals, includes in the binning process a weight coefficient for each pixel depending of its position in the spread out profile. This will drive a slight increase in the signal to noise ratio. This algorithm is described in the articles from K. Horne (1986) and J.C. Robertson (1986) and is in particular used by the IRIS software.

1.6 Wavelength calibration

By definition, each pixel of the spectrum profile contains the spectral intensity for a given wavelength. Up to now, we have the intensity by pixel graph, but the X-axis has not been graduated by wavelength. The relationship between the pixel in the horizontal axis *versus* the wavelength has to be determined. The pixel size, the spectral dispersion of the spectrograph will defined the wavelength range "seen" by a pixel. This is called the spectral sampling and is expressed in Å/pixel[1]. To establish the relationship pixel-wavelength, a calibration lamp is commonly used. This lamp is called a calibration lamp, as its known spectrum will help to establish the function. The lamp spectrum shall have easily identified lines whose accurate known wavelength will be used to calibrate the x-axis. The calibration exposure is done before and after the object spectrum exposure sequence. Doing it before and after could help to determine if any shift has occurred during the exposure. Calibration lamp can be from different type, but they shall fill some basic criteria:

- Lines well-known, and identified, be careful to exotic lamp which has some gas mix giving too complex spectrum.

- Emission lines.

- On the wavelength range covered by the sensor, shall have more than one line in this spectral range.

- Fine line, the finest they are, the more accurate will be the calibration.

In practical, if no lamp are available, a basic calibration can be considered using the spectrum of the object itself. In this case, if as for example, the Balmer lines of the star are used to calibrate the profile, it will not be possible further on to compute any Doppler radial shift of the star. One alternative still exists, which is acceptable. This alternative consists in not using the star lines but the earth atmospheric lines which are usually quite visible on the star spectrum itself as the star light had to cross the atmosphere. The wavelength of those atmospheric lines are well-known. They slightly vary in intensity depending of the pressure and temperature but their wavelength are stable. They also do not always exhibits a clean, symmetric shape, but can be used as a good alternate way or a complement if the calibration lamp does not have enough lines for an accurate non-linear calibration.

This method is extremely efficient to refine the calibration of a region centered on the H-alpha line by using the multiple H_2O lines which are superposed to the star spectrum. Once the accurate calibration process is done, those lines can be eliminated. We will see in a later section how to do this.

To establish the pixel-wavelength relation, several strategies can be considered depending of the desired accuracy and the mean available.

---

[1]Å: Angström, *i.e.* 10^{-10} m.

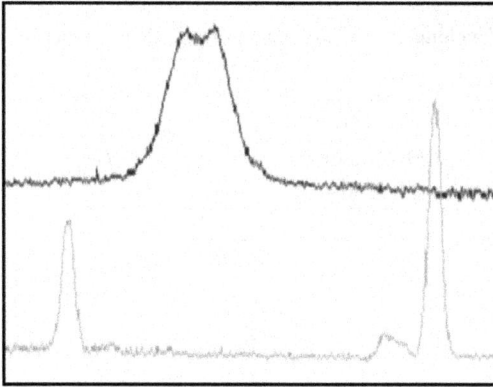

Fig. 17. Calibration spectrum of an Argon lamp superposed to the star spectrum – T60, Pic du Midi.

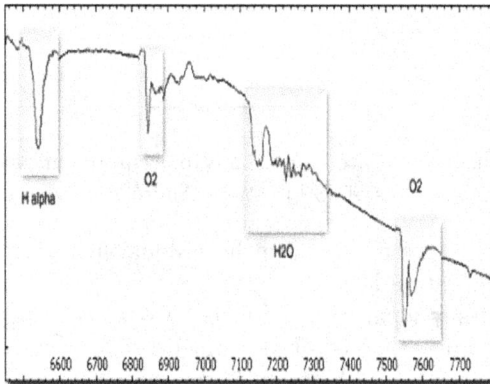

Fig. 18. Atmospheric lines in the near infra-red region – Vega spectrum.

If the sampling factor of the spectrograph is known, only one line whose wavelength is known can be used to define the offset. The dispersion will then considered to be constant across the wavelength range covered by the sensor, a simple linear interpolation. If the sampling factor is un-known, two lines with identified wavelength shall be available. The dispersion will again be considered as linear. In the reality, the dispersion is probably not perfectly linear. To establish the relationship, multiple lines will be required to determine the polynomial calibration. The relationship will interpolate from the multiple lines barycenter the best polynomial fit, with the degree limited by the number of lines provided.

In calibration process, a step-by-step approach can also be used. First step would be to make a linear calibration, then a multiple lines one. The linear one will help to more accurately determine the wavelength of the other calibration lines by increasing the degree of the polynomial fit. We usallly do not exceed the third degree polynomial fit. For this, four lines have to be identified. If more are

Table 1. Wavelength of some atmospheric lines in the near infra-red.

Wavelength (Å)	Spectral line
6575	O_2
≈ 7200 Molecular band	H_2O
7604	O_2

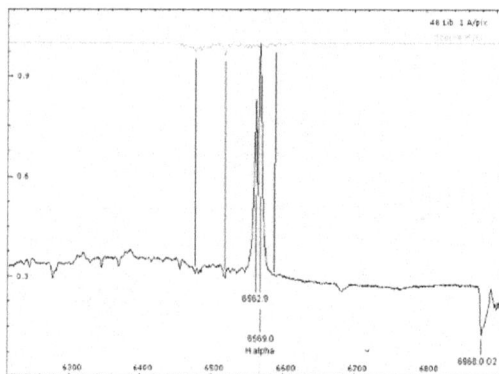

Fig. 19. Example of using the water vapor lines to accurately measure the doppler shift of the tow peaks of the H_α lines of 48 Lib star – André & Sylvain Rondi.

provided, they can be used to estimate the residual fit error, which even can be re-injected in the calibration law.

If lines are not easily identified, a strategy would be to first make a calibration with one or two lines, then check that the other lines correspond to expected wavelength. One the right combination is recognized, then all the lines can be used for the final calibration.

This calibration step is a crucial step for any scientific exploitation of the spectrum. It is recommended to validate the process on well-known starts, exhibiting few lines, easily identifiable, before jumping in the line identification of a complex spectral-type star. The first spectrum of any new comer in spectrography shall be Vega, or any bright A-type star, and not Arcturus or G,F,M type stars. Those complex late-type star have blended lines and the spectrum looks like a forest so very hard to identify with a limited sampling factor which line corresponds to which element wavelength.

2   Data reduction, final steps

This section described the final set of correction processing to get to the final "object only" astrophysical quantities:

- Spectral response correction

- Elimination of the telluric lines

- Heliocentric speed correction

- Atmospheric extinction correction.

2.1 Spectral response correction

As seen above in the flat-field section, the signal provided by the sensor is modified by its response or sensitivity at a given wavelength. The spectral response curve given by the constructors are just an indication and cannot be used to compensate the intensities measured. In addition, each sensor has its own response curve. To get back to the genuine signal of the object for each wavelength, the response curve shall be established. To define the sensor response curve on the observed wavelength domain, a spectrum of a reference star shall be made during the night. A reference star is a star from which its spectrum is available and is of course corrected from this effect. The comparison between the "true" spectrum and the acquired spectrum will give the response curve of the sensor. In this case, all the components spectral response are included, telescope, spectrograph optics, and not only the sensor. In the literature such reference spectra can be found. Some are calibrated in absolute flux or given in relative intensities (scaled at a given wavelength). Some catalog can even be download through internet. The use of a spectrum in absolute flux is only required if quantitative intensity measurement has to be done. If only relative intensities measurement are made, reference spectra in relative flux are recommended. The absolute flux computation would require the knowledge of the exposure duration, which is not need in the case of a relative intensity spectrum. In our case, what is important is to come back to the original shape of the continuum and not its absolute intensity value.

To correct the spectra from the instrumental response, the response will first be computed from a reference star. Let's take as for an example the Vega star, which is quite often used. In the following figure, the theoretical spectrum of an A0V type star has been superposed to the observed Vega spectrum. (Same spectral type than the observed star, spectrum from the database of A.J. Pickles 1998) – it is clearly noticeable that the observed spectrum profile stick to the response curve of a KAF400, low sensitivity in blue part, with a maximum in the visible red region.

The registered flux is equal to the real flux of the star, attenuated by the response of the instrument:

$$F_{obs} = F_r * R_\lambda. \tag{2}$$

As the real flux of the star is known, we can get to the attenuation for each wavelength (pixels):

$$R_\lambda = F_{obs}/F_r. \tag{3}$$

The simple division of the two spectra will unfortunately included the spectral lines. To get rid of them, and only keep the continuum variation modified by the response of the instrument, the resulting profile shall be filtered. A low-pass filter or a spline filtering can be used to eliminate the artifacts due to the presence of the spectral lines.

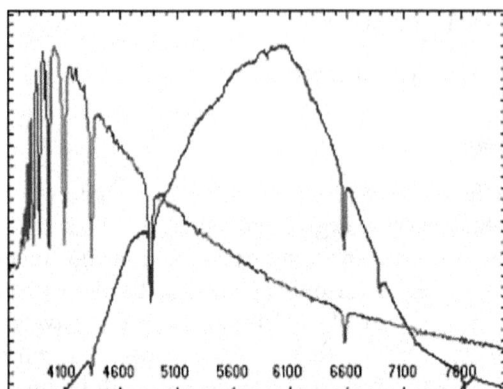

Fig. 20. Star observed spectrum, the maximum of the intensity is centered on the maximum of sensitivity of the sensor. The theoretical spectrum of the star is superposed for comparison.

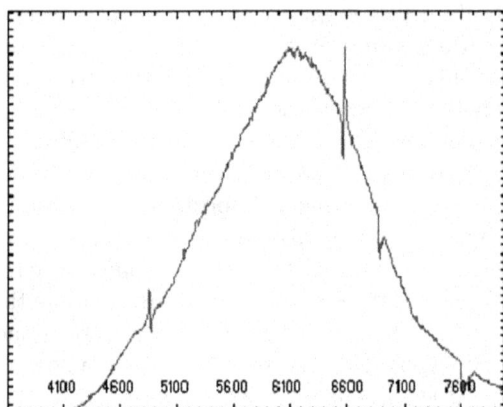

Fig. 21. Result of the division of the observed spectrum by the "theoretical" one of the star considered.

This graph represents the spectral response of the instrument. It can then further be used to correct all the spectra acquired during the night, by simply dividing them by this profile.

$$F_r = F_{obs}/R_\lambda. \tag{4}$$

This process is the last correction process which depends on the instrumental configuration. When exchanging spectra for specific studies among amateurs or with professional this is the last mandatory processing, as other user will not have access to your instrumental data. The forthcoming correction can be made by

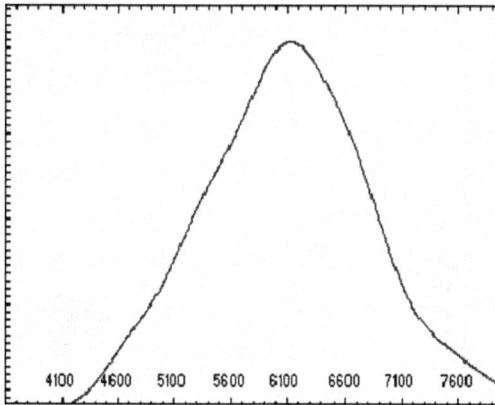

Fig. 22. After filtration, the division of the two spectra is the spectral response profile of the instrument.

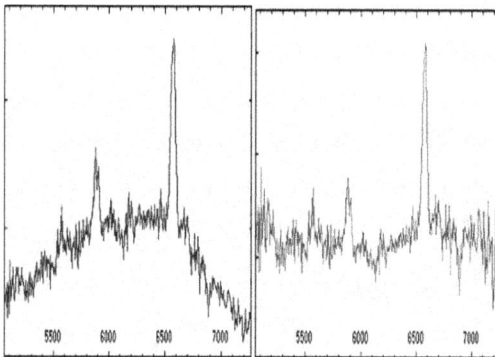

Fig. 23. EX Hya spectrum, cataclysmic variable, before and after the instrumental response correction.

anyone who knowns some basic facts about the observing site location, and the time and date of the exposure.

2.2 Telluric lines

In the section wavelength calibration, we have seen that spectral lines from the Earth atmosphere also called "telluric lines" can be very useful to calibrate accurately a profile. In this case, those lines are of certain use. But in fact and specially in medium and high resolution they clearly alterate the object spectrum by modifying locally their intensity, in another word they pollute the real spectrum of the object.

But, the telluric lines spectrum is perfectly known and can be modelized. A synthetic spectrum of the telluric lines can be generated, it then can be adapted

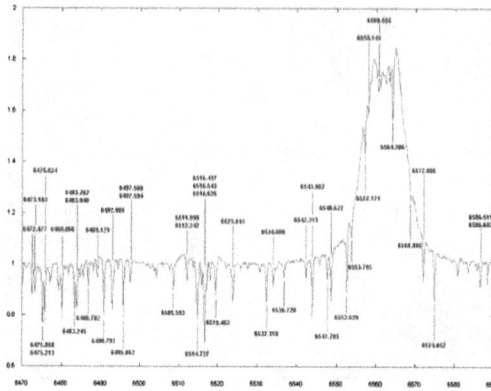

Fig. 24. Atmospheric lines around H-alpha region Sampling: 0.149 Å/pixel – 59 Cyg (4.392/08/1999).

Fig. 25. Telluric lines elimination in the H-alpha region. Vega spectrum.

with the parameters of the observing instrument: like resolution, attenuation. If the observed spectrum is then divided by this synthetic spectrum the telluric lines will be eliminated.

The telluric lines correction process only has a meaning if the resolution is sufficient to make them visible. It also requires that the spectrum is correctly calibrated in wavelength. If not, the division process will create artifacts by producing artificial lines, creating more defects than what we are trying to remove.

2.3 Heliocentric correction

To precisely measure spectral shift in wavelength, it has to be noted that the Doppler shift measured is always a relative one, being the difference on the line of

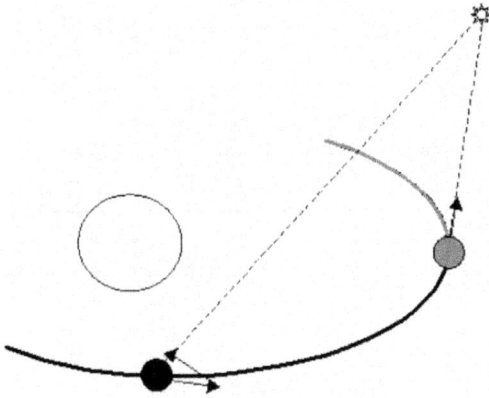

Fig. 26. Schematic of the Doppler effect produced by the Earth motion.

sight of the velocity between the observer and the object. In our case, the observer is on the Earth. And from one observation to another one, the earth has moved in the space around the sun (heliocentric motion), its motion vector can take several position all along the year, modulating the observed Doppler shift. By geometry, the motion vector can be calculated and this value is not negligible as long as medium resolution observations are performed.

The earth velocity in the direction of the object is a function of the observation date and of the position of the star on the celestial sphere at the time of the observation.

Earth motion equation are described as for example in the J. Meuus (1991) book and will not be repeated here. They utilized the following observation data:

- Date and time of the observation

- Observation site coordinates

- Object coordinates.

For the same object observed all along the year, it can be noted that the correction to be applied is cyclic, and has a period of one year, which is not surprising. If the Earth Doppler shift is not corrected, an error of the wavelength measurement of a spectral line will be made and the comparison of the same line profile at multiple dates can appear to be shifted.

As for example, if we consider the observation of the H-alpha line of a Be star. If the heliocentric correction is not made, the relative position of the spectral line will be shifted from one observation date to another. This shift is not a physical consequence of some variation of the star, but is due to the different position and motion vector of the Earth at the time of the observation. In the following figure, one could believe at a Doppler shift from the star, but once the heliocentric correction is applied, the spectral line are in fact back to be aligned.

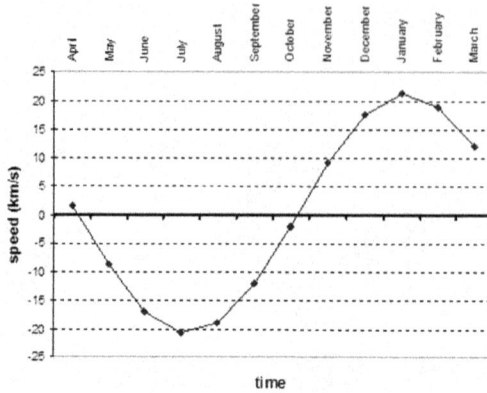

Fig. 27. Earth motion variation over a year period example – I CrB, Pic du Midi, T60.

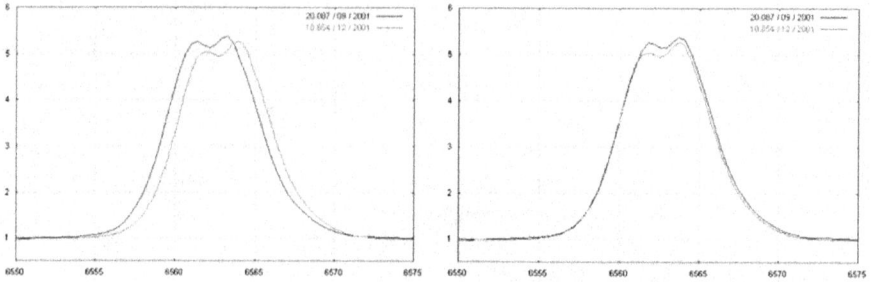

Fig. 28. H-alpha line position comparison for two observations separated by several months: not corrected and corrected from the heliocentric Doppler shift.

According to the doppler shift law, the shift in wavelength is given by:

$$\lambda_2 - \lambda_1 = (v_2 - v_1)/c, \tag{5}$$

with c the speed of the light. This represent a shift of 0.045 nm for a speed difference of $20 \, \mathrm{km \, s^{-1}}$ around the H-alpha line.

The relative Earth motion correction is thus required for:

- The comparison of observations of the same spectral line over several months

- The absolute measurement of the radial velocity of an object.

2.4 Atmospheric extinction

We have already seen in the telluric line section that the spectrum of a star is altered by the atmosphere. We considered at that time the superposition of the atmospheric spectrum. In addition to this, the atmosphere absorbs also a part

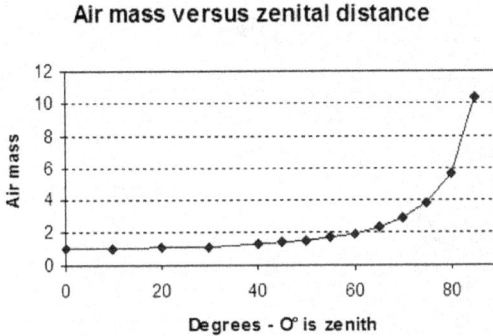

Air mass versus zenital distance

Fig. 29. Evolution of the air mass function of the zenith height.

of the light, depending of the object height in the sky and the of the wavelength considered. It is as if the atmosphere had its own spectral response, function of the air mass crossed by the object light. This attenuation can be calculated and then compensated to come back to the spectrum of the object as it is in space.

The attenuation of a monochromatic beam is, according to the Beer law, function of the thickness of the solution and its concentration.

$$LogI = logI_o - k_\lambda * M, \qquad (6)$$

with M the air mass crossed by the light.

As the coefficient k_λ depends of the wavelength λ, the attenuation will not be same on the overall range of the spectrum.

The air mass is the portion of the atmosphere crossed by the light on the line of sight at the time of the observation. It can be calculated knowing the site coordinates, the celestial coordinates of the observed object and the time of the observation determined by its hour angle:

$$1/M = sinh = sin\phi \cdot sin\delta + cos\phi * cos\delta * cosH, \qquad (7)$$

with ϕ = latitude of the observation site, δ = declination of the object and h = hour angle of the object.

The spectral absorbance of the Earth atmosphere can be explained by combination of multiple effects: the dispersion, so-called "of Rayleigh", the molecular absorption (presence of large molecular absorption bands) and the diffusion caused by the presence of dust particles (aerosols). Those processed modify the attenuation according to the wavelength of the radiation, and are also dependent of the atmospheric local conditions: pressure, altitude, water vapor, local pollution. For absolute flux measurements, the required accuracy would need that these effects are not corrected with the "standard atmosphere" simulation, but shall be based on comparative measurements using a reference star at the same zenith height, which *de facto* eliminate the need of the correction.

Table 2. Allen, Astrophysical Quantities, 3rd edition.

λ in nanometer (nm)	mag/air mass
400	0.24
450	0.16
500	0.12
550	0.11
600	0.1
650	0.06
700	0.04
750	0.03
800	0.02
850	0.02
900	0.02
950	0.01
1000	0.01
1200	0.0005

Professional observatories have establish their own model of their local atmosphere which simulate the extinction curve at given atmospheric conditions. This extinction curve is directly used by their reduction software to compensate an average extinction.

If the site extinction curve is not available, which is rarely the case in amateur world, a "standard" curve of extinction provided by Allen can be used. It has been calculated for a "standard atmosphere" (*cf.* Table 2).

The attenuation coefficient is expressed in magnitude by air mass. The absorption is much more important in the blue part of the spectrum than in the reed part.

In professional observatories, these coefficient are measured on tabulated range of wavelength. Some observatories publish their extinction curve and so, photometric conditions can be compared from one site to another one.

If some observatories like the La Palma's site is very close to the standard curve of Allen, some others have similar significant variations over the full wavelength range like the Cerro Tololo's site. Only the Mauna Kea site has a significant variation in the red part of the spectrum but with differences limited to 0.05 magnitude by air mass. We can consider that the usage of the Allen's curve is a good approximation if the general shape of the continuum is looked for.

For accurate flux measurement, the only robust way to proceed is to not to have to do this correction. The reference star for the spectral response curve shall be taken at the same zenith height than the observed object, thus the effect will be compensated, the beam of the reference star having crossed the same air mass.

For a spectrum which extends from 400 nm to 800 nm, we have seen that the atmospheric extinction is more important in the blue region than in the red region [*cf.* Table 2].

Observatories atmospheric extinction

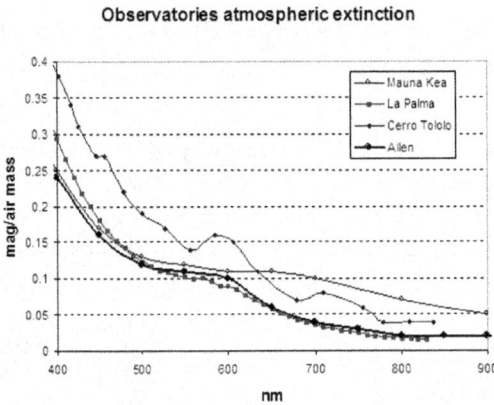

Fig. 30. Comparison of the atmospheric extinction for three professional observatories, see References Tololo C (1983), Mauna Kea (1988) and La Palma.

iota CrB

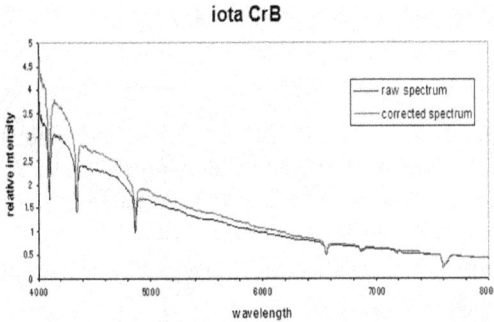

Fig. 31. I Crb spectrum corrected and not correction from the atmospheric extinction – Pic du Midi, T60.

Based on the observation of the star ICrB made at the Pic du Midi the 24th of April 2003, applying the extinction correction will effectively modify the intensities of the continuum in the blue part of the profile. As the continuum shape is sometimes used to determine the star temperature with the application of the Planck law, the extinction correction will improve the estimation.

Quantitatively, how this effect is important? if we consider a star at a zenith height of $45°$ – the estimated air mass is 1.4 – for a wavelength of 550 nm, the attenuation given by the Allen's table is 0.11 mag/airmass, which gives a total value of 0.15 mag. To calculate the relative intensity error, we can use the formula which links the magnitude variation $M2 - M1$ to the flux $F2$ and $F1$, here:

$$M1 - M2 = -2.5 \, log \, F1/F2 \qquad (8)$$

Reverting the formula,

$$F1/F2 = 10^{-Delta \, mag/2.5} \qquad (9)$$

showing that the error in the intensity ratio (or flux) for an atmospheric extinction of 0.15 mag will thus be of 0.94, which is a variation of 12% for wavelength around 550 nm.

If a study is conducted on the evolution of the H-alpha line, this correction has insignificant effect. Not only the H-alpha line is located in the red part of the spectrum where the extinction effect is not too important, in the order of 0.05 mag/airmass, but in addition the attenuation coefficient slowly varies on a such reduced wavelength range. We thus can skip this correction and keep it only for specific studies on large spectral range of the continuum, in the blue part of the spectrum.

We also have to notice that an error in the data like the date of the observation time will drive an erroneous correction. And, as mentioned in the atmosphere model paragraph, this is only an estimation. The more robust method is to avoid to have to do it, by picking a reference star at the same zenith height than the observed object.

3   Measurements and astrophysical tools

3.1   Measurements

Once the spectral data being reduced and the global star spectrum obtained, some may want to go further in the analysis. The simple comparison of spectra does not give access to quantitative analysis. In this section, we will describe some measurement method of quantitative parameters on line profile and the usage of some tools which can help to make further steps in the analysis.

Line center. The determination of the center of a spectral line is very often used in the numerous steps of the data reduction or in the final production of scientific measurement. The computation accuracy of the line center is nevertheless a tricky process.

In the calibration process, it is the association of the line center with the theoretical wavelength which will drive the the accuracy of the relationship pixel-wavelength. In the analysis phase, the line center measurement will be done to compute a relative Doppler shift or help to remove some ambiguity on a line identification.

The line center cannot by any mean be assimilated to its maximum. Multiple physical processes and instrumental ones can modify the line profile. The line can be asymmetric, can have multiple maximum, and be largely disturbed by the evolution of the continuum.

The following method shall be considered:

- Simulation of the line profile by a function of Gaussian type, then analytical computation of the line center, and width at mid-height. The limitation of this method resides in the fact that profiles does not always follows a Gaussian shape, and this also requires that the continuum is flat around

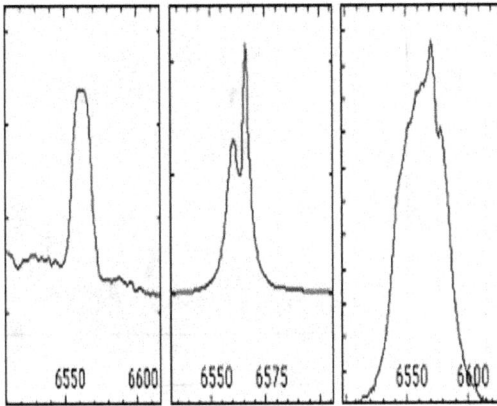

Fig. 32. Different examples of line spectral profile: 60 Cyg, continuum variation – Dz Tau, multiple maxima – Nova Cyg 1992, multiple lines.

the spectral line. Of course, some other function can be considered: Lorentz profile, Voigt profile, but the assumption of the symmetry of the line is still a constraint, which if not matched is a source of error. – Barycenter computation: this method is more generic and robust, can apply to asymmetric lines. One drawback is its sensitiveness to the definition of the line boundaries but is quite easy to implement.

What ever the method used, one tricky point is to define the boundaries of the line shape to considered. Very often this step is a manual one, and is so a subjective process which drives to accuracy limitation. On top of this, some small lines can be captured. Including them will modify the line area calculation. Other source of errors, a continuum having a strong slope around the line will also alterate the accuracy of the barycenter.

Finally, a too narrowed selection, would also be damageable as it would include only partially the energy of the line, excluding the one in the line aisles.

The following study shows the evolution of the width at mid-height of the H-alpha line of Vega for at medium resolution for different boundaries selection of the line profile.

Attention shall be paid when line boundaries selection has to be made, avoiding to include close by lines and not too narrow excluding the energy contained in the line aisles.

Equivalent width. The equivalent width represent the strength of a spectral line, meaning the energy emitted or absorbed in the spectral line. From a scaled to one continuum around the line itself, this width is defined as the surface of a rectangle which its surface will be equal to line surface above the continuum C_λ and its profile F_λ. The equivalent width is given by:

$$W_{eq} = \int (C_\lambda - F_\lambda)/C_\lambda d\lambda. \tag{10}$$

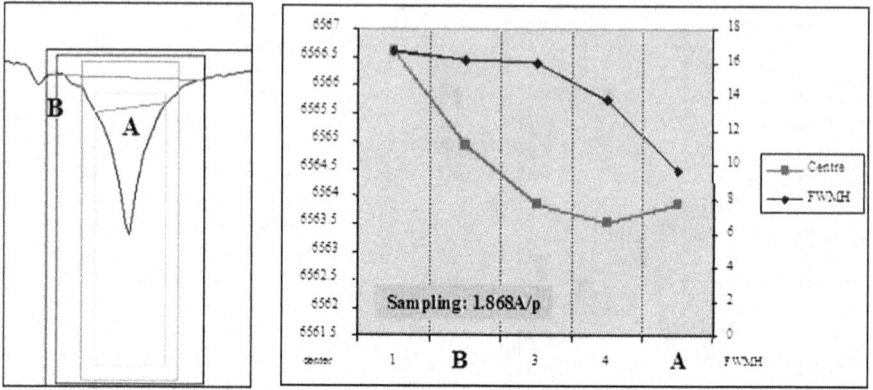

Fig. 33. Impact of different selection of the boundaries on the computation of the line center and its width at mid-height.

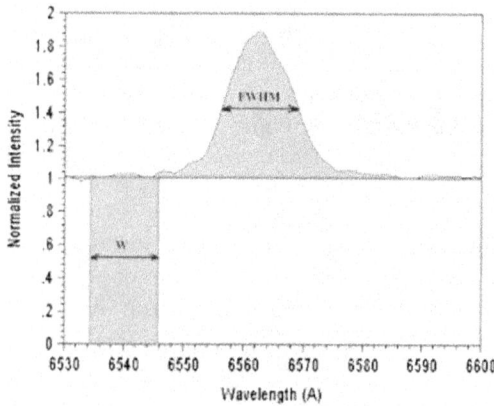

Fig. 34. Equivalent width definition of a spectral line.

By convention an equivalent width of an absorption line will be negative, and positive for an emission line.

Among all the regular studies involving the computation of the Equivalent Width, one could be the monitoring over time of its value for a given spectral line of the same object. The interpretation of the data has however to be cautious. The apparent strength of the line can in fact be altered by the evolution of the continuum itself. If the EW decrease, it can be a real decrease of the line energy or the intensity increase of the continuum around. A sudden variation of the continuum can altered the interpretation of the equivalent width evolution. To remove the ambiguity, a photometric measurement will be necessary to complete the analysis in the case of sudden variation (like in nova observations). If the star is not suspicious to be a variable star, then in this case the interpretation is valid.

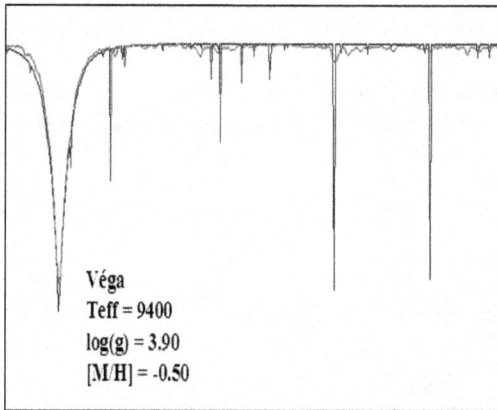

Fig. 35. Vega spectrum synthetic spectrum at 0.1 Å/pixel and observed spectrum at a sampling of 1.41 Å/pixels – Spectrum software from R.O. Gray.

3.2 Spectral lines database

It is quite useful to have spectral line tables available, tables established in laboratories, describing with precision the nature and the the wavelength of the atomic lines. The principal source of reference is the CRC Handbook of Chemistry & Physics, which can be found on the web (CDS catalog, 1980–81).

Some lines catalog, specific to a type of study can also be found in the literature. They represent a sub-set of the all atomic lines, listing the most probable lines to be found in the spectra considered. As for example, the stellar lines catalog of (Coluzzi 1993–1999).

Some software can embedded such database to ease the spectral line identification or wavelength entering. Professional observatories also have specific tables, which contains only the lines of the calibration lamp used by the spectrograph.

3.3 Stellar spectral synthesis

Spectral lines formation involves physical processes which are more and more understood. The equation can be embedded in a simulation program and some of them can be found in the literature or on the web. These programs can go up to the complete synthesis of a star spectrum, using only a limited number of physical parameters to describe the star. The R.L. Kurucz (1993) models are the more largely cited among several others. Based on a stellar model corresponding to a given effective temperature, a gravitational pressure and a micro-turbulence value, a stellar spectral synthesis program like "Spectrum" from R.O. Gray generates the theoretical spectrum of the star considered. The theoretical one and the observed one can be compared. The parameters which describe the star model can be changed until the two profile matches and thus derived from here the real star physical parameters.

Those very powerful tool are based on a good knowledge of the physic and an understanding of the intrinsic limitation of the modelization used. A careful usage must be made of them, specifically without a real knowledge of the physics limitation. The combination of those synthesis software tools with the availability of numerous database makes us foresee that more and more tools are now available and accessible to the amateurs.

4 Conclusion

Spectrograph construction adapted to the specific study of certain objects is of course a very important step in spectroscopic observations. If the data acquisition is the next mandatory step, the data reduction is essential to complete the observations. A spectrum not calibrated in wavelength is like an image not fully processed. We have seen above some corrections or calibration which are clearly mandatory before exchanging spectra or comparing them, and some which are more dedicated to a final reduction into fundamental astrophysical parameters.

This final reduction step shall be done with the same rigor and cautious than the construction or acquisition. One error in manipulating the data, misscalibrating the profile or creating false lines will definitively drive erroneous physical interpretation. This step is mandatory before sharing the data with professionals. And we will see with the example of the Be stars study chapter from Coralie Neiner and with those of Nicolas Biver, that amateur data may be very useful for professional studies.

The correction step involving the bias, thermal signal, flat-field, registration, profile reduction, wavelength calibration and spectral response correction are the minimum processing to be applied before sharing the spectrum with community. At each step, the informations can be degraded and lost which is a pity considering the effort made to capture each photon.

If a professional collaboration takes place, the atmospheric, heliocentric or quantitative measurement on the profiles can be made by the professional with its own favorite tools and a consistent process applied across all the data collected from other source. If on the other hand the amateur wants to go him/herself at the discovery of the astrophysical quantities, some explanation in this chapter may lit his/her path through the available methods, quantities and tools available and their application domain.

It is quite important than the amateur masters all these steps. The credibility of the work is at stake and it is a sine qua non condition to consider a stable and enriching collaboration.

References

Pickles, A.J., 1998, PASP, 110, 863

Horne, K. A.S.P. An Optimal Extraction Algorithm for CCD Spectroscopy, 609–617, Pub., 98, 1986

Robertson, J.C., 1986, PASP, 98, 1220

Meuus, J., 1991, "Astronomical Algorithms" (Willmann-Bell, Richmond)

Tololo, C., Stones, & Baldwin, 1983

Mauna Kea, Boulade, 1988, CFHT Bulletin, 19, 16, for l < 400 nm

La Palma, Isaac Newton Group of telescope - Roque de los Muchachos - Observing guide:
    http://www.ing.iac.es/Astronomy/observing/manuals/html_manuals/general/
    obs_guide/node293.html

CDS catalog: VI/16 Line Spectra of the Elements (Reader+ 1980–1981) - Reader J. &
    Corliss Ch.H., 61st ed., CRC Handbook of Chemistry & Physics (1980–81)

CDS catalog: VI/71A Revised version of the ILLSS Catalogue (Coluzzi 1993–1999) -
    Coluzzi R: 1993, Bull. Inf. CDS, 43, 7

Kurucz, R.L., 1993, ATLAS9, SAO, (Cambridge, USA) http://kurucz.harvard.edu/

Gray, R.O. Department of Physics and Astronomy, Appalachian State University
    http://www1.appstate.edu/dept/physics/spectrum/spectrum.html

Astronomical Spectrography for Amateurs
J.-P. Rozelot and C. Neiner (eds)
EAS Publications Series, **47** (2011) 103–137

THE SUN'S SPECTRA: CODING THE LIGHT AND SOUNDS. APPLICATION TO OTHER STARS

J.-P. Rozelot[1]

Abstract. The Sun is our nearest star. Its thorough study permits to extend results to other stars for which one does not think at once to encounter features first discovered on the Sun: spots, differential rotation, oblateness, radial surface displacements, etc. The Sun is thus an irreplaceable laboratory, as much as physical conditions prevailing there are often hardly reproducible on Earth. In this chapter, we do not intend to give an exhaustive survey of what we know about our Sun. We want only to give an original lighting on topical questions related to the subject of this book, based on stellar spectroscopy, and we will focus on what we call *the solar code*. What can we learn from the solar spectrum? More generally, what are we learning from solar oscillations and from the activity cycle? This chapter is divided into three parts, bearing in mind that results obtained on the Sun are transferable to other stars.

In a first part, we will show that the electromagnetic light code permits to access to different atmospheric solar layers. In a second part, we will show that the sound code is a fantastic tool for investigating the internal structure and the dynamics of the Sun. Thus tackled, the Sun is "peeled" as an onion, each successive shells giving an indication on the physical conditions acting in the studied layer, starting from the external atmosphere, crossing the free surface, to progressively go deeper inside, down to the core. In the third part, we will emphasize the "shape" concept, as departures to sphericity are essential in such an approach. This also allows us to study some global astrophysical properties, such as the angular momentum, the gravitational moments and the effect of distortion induced on the visible surface. We will conclude by extending such ideas to other stars, and especially by mentioning new results obtained on the oblateness of Altair and Achernar, including gravity darkening and geometrical distortion.

We intentionally replaced this whole matter in an instrumental context, by highlighting the observations, to the detriment of mathematical formulations, often tempting, but difficult: the reader will be able to find them, thanks to the many bibliographical references given.

[1] Université de Nice-Sophia Antipolis, Fizeau Dpt, CNRS UMR 6525 & Observatoire de la Côte d'Azur, Avenue Copernic, 06130 Grasse; e-mail: `rozelot@obs-azur.fr`

DOI: 10.1051/eas/1147004

1　What is the Sun?

The Sun is a fluid body in slowly non uniform rotation, whose surface manifestations may cause effects up to the level of the Earth and even beyond. They vanish at an ultimate frontier, called *heliopause*, which determines the gravitational, radiative and magnetic sphere of influence of the Sun. These three components characterize our star from a physical point of view: the study of the distribution of the solar mass and its luminosity, coupled to its magnetic field determine (almost) completely the state of the Sun.

Indeed, the distribution of mass determines the external and internal gravitational potential of the Sun; the luminosity (L) is connected to the temperature (T) of the star, at least to first order, by means of the Stephan's law ($L = \sigma T^4$); moreover, the electric or neutral particles that determine the composition of the star are transported along the lines of force of the magnetic field created by the non uniform rotation of the fluid. This physics characterizes almost a quiet Sun; to this one are superposed periods of activity, more or less cyclic, more or less intense. Sometimes other violently eruptive phenomena occur, which are for the moment still unpredictable, ejecting in space important quantities of matter, known under the name of CME: *"Coronal Mass Ejection"*.

Strictly speaking, it would be necessary to distinguish between two Suns: the Sun as it is really, and a quiet Sun, to tell the true difficult to observe, but whose properties would allow to define a mean state that would characterize various fundamental parameters. For example, one could define the quiet Sun by the mean particle density at a given altitude, in a quasi spherical symmetry. One could say also that the quiet Sun is characterized by a mean background of known brightness, etc. On this quasi ideal Sun are superposed two effects: one is a modulation that takes the form of periodic waves. For example, the oscillations in intensity of the surface, of well regular periodicities, or the oscillations of the magnetic field, more pseudo-periodicals. The other effect is a real cataclysm, whose violence disputes to the unpredictability, but whose effects at the level of the terrestrial orbit can be devastating, and that one analyze nowadays under the vocable of *"space weather"* (Rozelot 2006).

The reader will find in many books and manuals the totality of fundamental parameters of the Sun and the manner they were obtained. We have not judged useful to reproduce them here. This chapter forming a part of courses delivered in a CNRS school dedicated to the spectroscopic observation of stars, we have sought to give to the amateur an original vision of the current states of the researches. Certainly, the professional astronomer will also find there matter of thinking.

The spectroscopy being classically "the art forcing the light to speak", we have widened the concept to the analysis of the sounds. We will give therefore in this chapter a broader meaning, more physical, and we will attach ourselves to find a kind of genetic coding of the Sun, as if we had to clone it later on...

2　The solar spectrum

The Sun emits in all wavelengths, ranging from the gamma radiation at very short wavelengths (of the order of 0.1 nm), up to the most far radiation, at very long

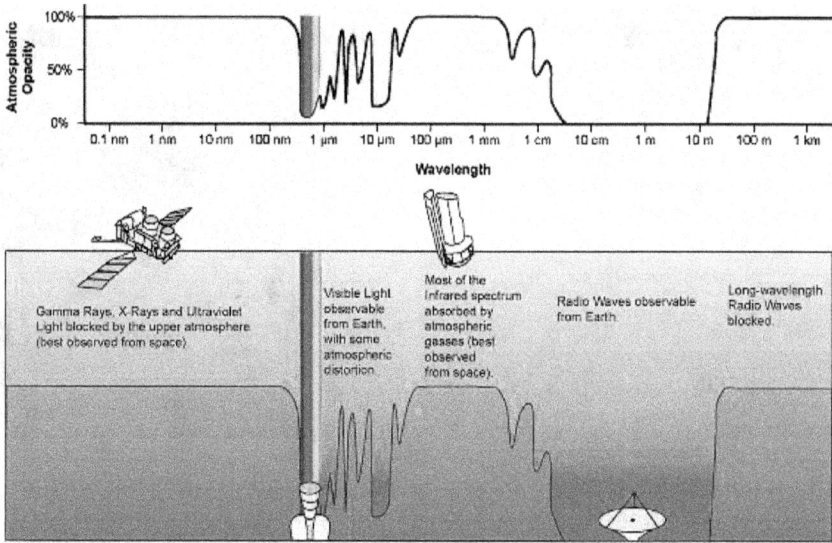

Fig. 1. Radiation spectrum range emitted by the Sun and "windows" of the terrestrial atmosphere: these regions are the only one which allow the radiation coming from the star to go trough our atmosphere and are therefore accessible from the ground. The two fields called "visible" and "radio" are thus the two alone which permit observations.

wavelengths, of the order of several kilometers. Figure 1 visualizes this extended range of frequency. However, the terrestrial atmosphere is more or less opaque, and let filter only some specific waves called visible and radio. Frontiers are not frozen, because they depend on the place and the altitude of the site where they are observed. For example, it is commonly admit that the UV limit of the solar spectrum is 320 nm, what is usually the case in lowland when the Sun culminates to its zenith. But this limit lowers to 290 nm if one observes at 5000 m above the sea level.

In practice, a light beam emitted by the Sun travels indefinitely, as long as it will cross the space without meeting an obstacle, atoms, electrons or other solid bodies (for example space debris from human manufacturing!), against which a reflection or an absorption will be made. If no shocks occur, its message can be transmitted until the limits of the Universe. When a wave encounters some matter, the photons are absorbed; by contrast, they are created when the matter emits a radiation, for example when the matter is heated. However, a given atom can absorb or radiate only a very specific set of photons of well determined energy. The adopted unit is the electron Volt (eV) that corresponds to an energy acquired by a particle of equal charge to that of an electron, accelerated by a difference of potential of 1 Volt. A photon of visible light has an energy of the order of 2 eV. The energy of the photon is conversely proportional to the wavelength (therefore proportional to the frequency). A short wavelength radiation corresponds to photons of great energy, so that the astrophysicist rather prefers to speak in terms of energy when he is addressing to radiations of very high frequencies.

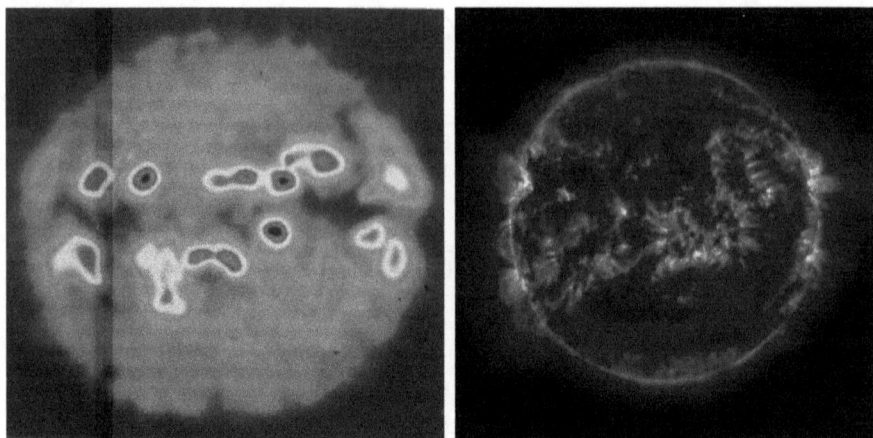

Fig. 2. *Left*: the Sun as it appears in radio waves. Image obtained on 26 September 1981, at 20 cm of wavelength with the VLA (Very Large Array), situated at Soccoro, New Mexico (USA). *Right*: the Sun observed on 24 July 2002, in the Fe X emission line ($\lambda = 171$ Ångströms), by means of the EIT instrument embarked on board of the SOHO spacecraft. In the two cases, the photosphere, too cold, does not emit, and is not visible.

Figure 2 (on the left) shows the Sun observed at a wavelength of 20 cm. The shape of the disc seems blurred at a first glance, but reflects simply the density and temperature conditions reigning at the altitude where the radiation is emitted. The selection of a well defined wavelength within the solar spectrum, allows to focus on a specific layer of the solar atmosphere at a well determined altitude. When observing the Sun outer the terrestrial atmosphere, by means of satellites, one has access to the totality of the emitted electromagnetic spectrum, therefore it is possible to detect layers more or less deeper anchored in the solar atmosphere. For example, in the extreme UV radiation, the photons are very energetics and are emitted only in a very hot envelope all around the Sun, called the **corona**. The solar disc itself is far more cold, therefore does not emit at this wavelength and remains invisible. In X radiation, the corona is thus visible by transparency on the totality of the disc with a very high resolution (Fig. 2, right, obtained at $\lambda = 171$ Ångströms).

The case is similar in the wavelength radio range: the photosphere does not emit in this range of wavelength, and is not visible. On the other hand, very hot zones, in general plasma trapped in magnetic loops developed above solar active centers, emit intensely in radio. These localized structures can be seen on Figure 2 (left).

In the visible field, we receive on the Earth the radiations coming from the totality of the visible envelope of the Sun. Radiations emitted at the "periphery" are emitted in a thin layer, and the more one looks to the center of the disc, the more the quantity of matter crossed is important. Thus one conceives intuitively that when going gradually from the outer to the center, the constitutive gas of the

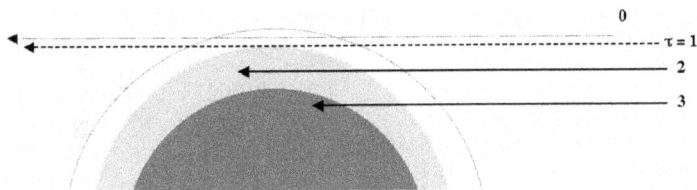

Fig. 3. The optical depth concept. Rays labeled 0, 1, 2 and 3 simulate three lines of sight; the gray part shows the different solar layers, more opaque with growing increase to the interior. The ray "0" crosses an extremely tenuous layer, and "sees" the totality of the atmosphere. Rays "2" and "3" penetrate less and less deeply in the solar atmosphere, along the line of sight, because of the opacity; the ray "1" just reaches the limit between the transparent layer and the opaque layer. We say that the *optical depth* τ is 1, which defines by convention the border of the photosphere.

Sun will pass gradually from transparent to opaque. The layer where this change takes place, defines the **photosphere**: it spreads on some hundred of kilometers only (≈ 300 km). In practice, all the information we get on the Sun in visible light comes from this layer (Fig. 3).

In definitive, the electromagnetic spectrum gives informations on the physical states of temperature, density, magnetic field, which occur to a given altitude of the gaseous solar "sphere".

2.1 The "integrated" spectrum

By cutting the solar spectrum in small slices of wavelength, we may access to a particular region of the Sun. To take an image, we may assimilate the Sun to an onion; it is thus as if one "focus" with a binocular on such or such peel, including a "defect" on a local place. Each slices of the spectrum is accessible with a particular technique, the most common one being the *spectrograph*[1].

Figure 4 shows the different layers probed by means of various indicators of the electromagnetic radiation, going gradually with increasing altitude, from the photosphere to the **chromosphere**, then, from the **transition zone** where the temperature of the medium increases abruptly from some 10 000 K to several millions K, subsequently to the **corona**, first the lower corona, then the middle corona and finally the upper corona, this last one being well accessible to radio waves.

The total integrated spectrum of the Sun leads to an important notion for the understanding of the global machinery of our star, and has aroused a renewal of interest since a twenty of years. From the EUV to the visible, the Sun radiates continuously, but the average background shows large variations with the wavelength (Figs. 5 and 6). This background culminates in the visible, then decreases in the IR part of the spectrum. This last part follows, approximately from $\lambda = 500$ nm

---

[1]Other instruments such as polarimeters or magnetographs, which are less accessible to the amateur, will not be described here (even so see paragraph 4.)

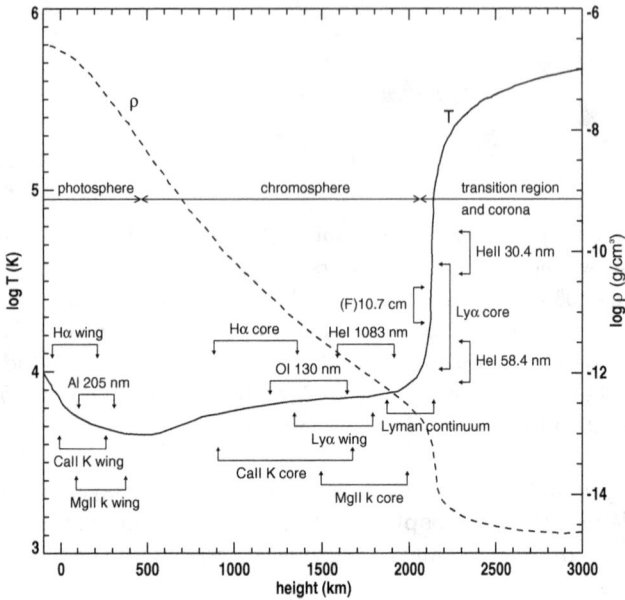

Fig. 4. Probe of solar atmospheric layers by means of various indicators of the electromagnetic radiation. Note the curve of temperature T, with its minimum (continuous line, left scale), and the curve of density ρ (dots, right scale). The notation (F) stands for the radio flux (at 10.7 cm). Note the minimum of temperature which occurs at around 500 km. According to Floyd (2003).

to 100 000 nm, with an astonishing precision, the radiation of the Planck law[2] for the red wing, which is ($h\nu \ll kT$):

$$S(\lambda) = 2ckT\lambda^{-4}. \tag{1}$$

By contrast, the UV part shows a more complex variability, both in amplitude and in time, and its influence on our climate system is suspected since a long time, as we will show a bit latter on. Let us recall that irradiance is the electromagnetic energy received by unit of time and unit area outside the atmosphere. The in-orbit measurements being made days after days, one has to carry back this quantity to 1 UA, to be free from the eccentricity of the orbit of the Earth. We must distinguish between such irradiance and the **total irradiance** (Total Solar Irradiance – TSI), that is the integral (in a mathematical sense) of the electromagnetic flux over all the wavelengths. Mg I and Mg II lines, respectively at 278.6 nm and 280.1 nm, have appeared to be extremely sensitive indicators of the solar activity. They are now included in the set of the perfect user of solar data (see Fig. 9).

The total solar irradiance variability with time has also been put in evidence since a few years (two decades). Figure 7 gives a composite setting (this means that

---

[2]With usual notations, ν the frequency, c, the light speed and k, the Boltzman constant.

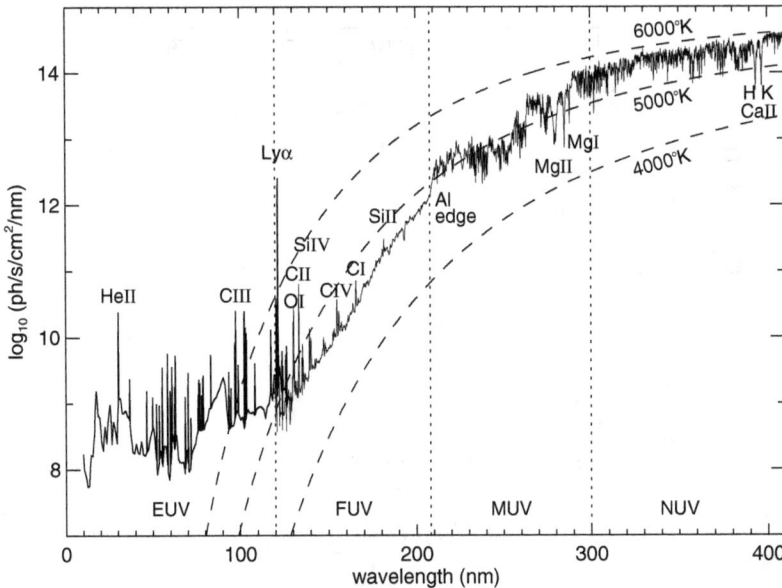

Fig. 5. Spectrum of the solar radiation from the EUV up to the visible. The average background grows regularly, streaked by numerous emission lines produced by the various constituents of the medium. This average level shows a very great variability with the solar cycle, especially in the EUV, and more especially in the Mg I and Mg II lines. In abscissa, letters E, F, M and N stands for Extreme, Far, Mean and Near –UV– (According to Floyd 2003).

the different measurements coming from different satellites have been rescaled on a unique "absolute" scale) of the TSI since the space measurements commencement. The variability reaches 3 per 1000 from a minimum to an other one, peak to peak, but does not exceed 1 for 1000 when quasi daily measurements are averaged over a month. Thus, as already mentioned by T. Moreux (a French Abbot –1867–1954– impassioned observer of sunspots), in its book on *"The Sky and the Universe"*, *1928, Doin ed., p. 24*: "It is this value that was called formerly rather wrongly the solar constant, and as I have noticed well before American scientists did it, this constant is extremely variable and should have been named the solar inconstant"[3].

Large efforts have been deployed these last years for modeling the total irradiance, and several approaches have been proposed. The most frequently used consists in calculating a pondered function of the area of spots and faculae. The darkening due to spots comes in diminution of an average bright and calm background, and the "embrillancement" –*i.e.* shining enhancement– of the faculae acts as an increase. This process proposed for the time first by P. Foukal (1979)

---

[3]Formerly TSI was called "solar constant". To avoid any confusion, the term "solar constant" should be used today only to describe the long-term average TSI.

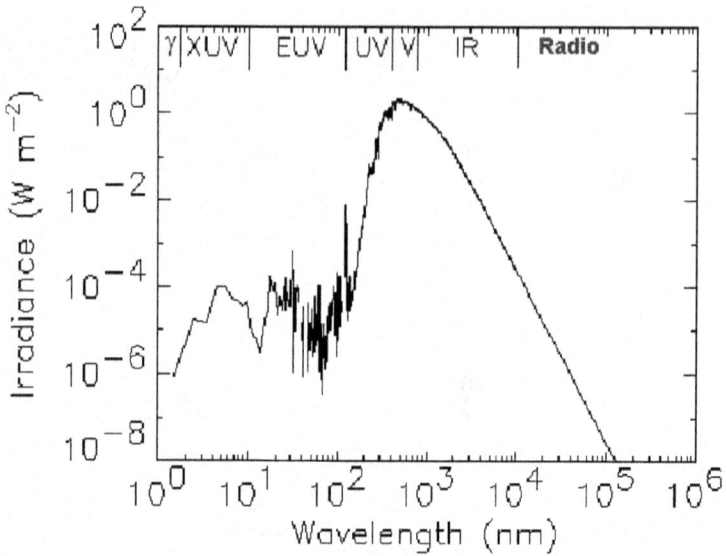

Fig. 6. Entire spectrum of the solar irradiance plotted as a function of wavelength. On the right, the straight part of the curve represents the IR trend, and follows the well known Planck's law, *cf.* Equation (1). On the left, the variability of the UV part is well noticeable. Figure 5 shows a zoom of this part.

has been taken again several times by other scientists, and has been improved since then, giving very good results, as the total irradiance can be modeled in such a way to approximately 94 per cent. During solar maximum periods of time, there are more sunspots, but the faculae are more intense, *i.e.* the deficit of darkness due to spots is not compensated, and so the TSI is higher than during periods of solar minimum. In this approach, the underlying physics supposes that the irradiance reflects simply the surface magnetism. As spots and faculae indicators are known since a long time, this process allows to reconstruct the irradiance over past times. In another suitable process intervene global parameters of the Sun, mainly the magnetic field, the Sun's radius and the efficient temperature (Sofia & Li 2001). The irradiance is then the result of stellar structural parameters connected between them by physical laws. This approach is certainly more convincing, albeit more difficult to implement. A 3-D model has been recently proposed which permit to sharpen our knowledge on the solar core (computations made by S. Mathis and S. Brun at the CEA in Saclay –F–).

In spite of some authors still reluctant, the study of the **solar irradiance** has become an entire part for studying the terrestrial climatic system. Indeed, the UV part of the spectrum is very time dependent. It is a quite recent discovery, that could have been believed as a pure fundamental research result, but appears to have important consequences on the climate of the Earth. This temporal variability is illustrated in the left part of Figure 8. In the visible part of the spectrum,

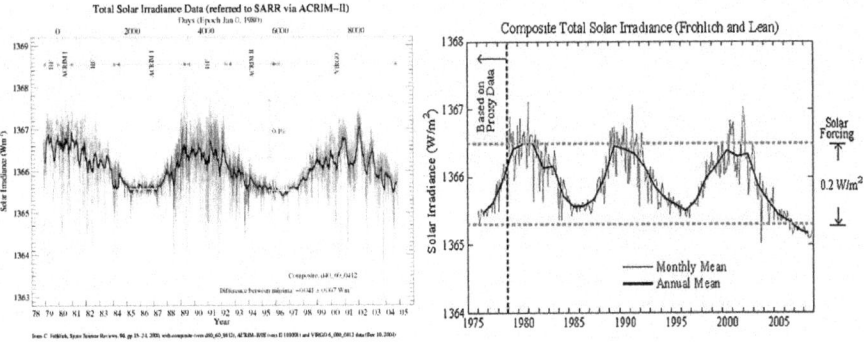

Fig. 7. *Left*: total Solar Irradiance (TSI –area below the curve of Figure 6 observed since the beginning of space measurements era, as a function of time. Measurements cover a whole cycle (22) and two half-cycles on each sides. (According to Frölich 2002). Measurements come from different satellites (each color represents data obtained from a dedicated satellite over its life-time period), and has been rescaled on a common scale. *Right*: the same curve extrapolated to cover cycle 21 and extended to the end of 2008. (According to Frölich & Lean 2010). It can be seen that in 2008, the TSI intensity was the lowest recorded.

the ratio between the irradiance estimate during the solar activity maximum period of time and its estimate during the solar activity minimum is of the order of 1. This ratio can reach 2 around 130 nm. The EUV flux acts by photodissociation of the particles, notably the oxygen molecule (O_2) and the ozone one (O_3), which are the major constituents of the upper atmosphere (Habereiter 2002). The right-hand curve shows the penetration depth of the solar radiation in the atmosphere, in the presence of various chemical species. It is clearly seen that the EUV solar irradiance variability may play a major role in the upper atmosphere, much less in the lower atmosphere (10–20 km), and in the near ultraviolet range. Therefore, it is the whole chemical composition, and particularly the ozone contained in the high stratosphere, which is sensible to the global change of the solar irradiance. At a lower altitude, ozone will be the key species for the coupling and feedbacks of the system chemistry/climate. The denominations (Herzberg, Schumann – Runge, etc.) indicate spectral bands for which studies were made in laboratory. These studies show that over the whole range of wavelengths, the values of the cross section in these continuum are much smaller than the values used in many photochemical stratospheric models, but in agreement with stratospheric in-situ measurements.

2.2 Solar spectra accessible by the amateur

For the amateur, and at a first glance, it seems difficult to get observations from space. However CNES (or other space Agencies) announces regularly calls of tender, aiming to propose not too complicated experiences being able to be embarked

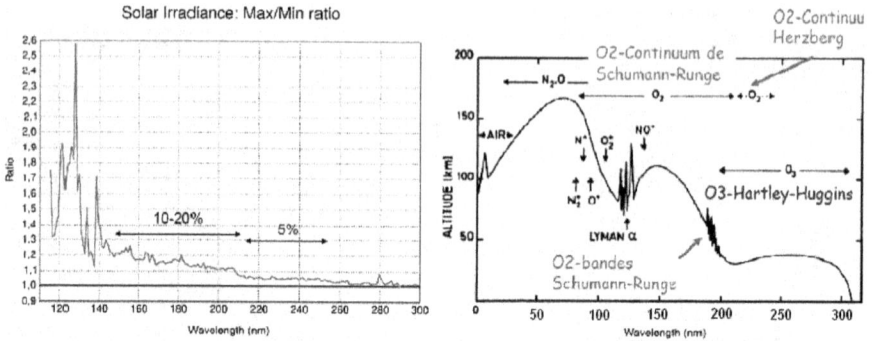

Fig. 8. *Left*: temporal variations of the total solar irradiance. In ordinates the ratio (value obtained during the maximum of the cycle)/(value obtained during the minimum of the cycle) has been reported, and in abscissa the wavelength. One can see that in the visible, this variation is very low, but can reach a factor 2 and more around 130 nm. *Right*: penetration depth of the solar radiation in the atmosphere. The dissociation of the chemical elements of the very upper atmosphere is more significant at very short wavelengths, showing the influence of the solar activity. (According to Marchand & Hauchecorne 2010.) The named continuum indicate spectral bands for which the contributions of UV photodissociation were more especially studied.

on board of rockets as micro-satellites. Such channels could be more exploited. From the ground, two fields of research can be approached: the radio and the visible domain.

2.2.1   The radio domain

This field is in general scarcely approached, because it has been perceived for a long time, either difficult or expensive. Today, these two reasons are largely overshoot. For an enthusiastic amateur, it is relatively easy to construct the totality of a radio waves receiver, for example at 100 MHz, *i.e.* a wavelength of 3 cm. It is sufficient to get an antenna, such as a typical TV satellite dish, and a decoder-recorder to store the signal on a computer. However, the amateur will be quickly confronted to a difficulty, just as professionals, which is the resolution. Indeed, the resolution of a radio antenna (as any other telescope mirror) is given by:

$$R = 1,22\lambda/D$$

where λ is the wavelength and D the diameter of the antenna. For example, for $\lambda = 3$ cm and $D = 100$ cm (a dish of this size can be easily found in specialized shops), the resolution is $3.66\,10^{-2}$ radians, that is to say 126 arc minutes, 4 times more than the Sun diameter (32 minutes of arc). One can see that the Sun is unsolved, and the inverse calculation shows that it would be necessary to have a 400 cm antenna of diameter to observe the whole Sun, and approximately a 400 m one to begin to perceive details, a hardly acceptable dimension for the amateur.

On the other hand, it is the reason for which such dimensions of antennas are found, at Nançay (F) for example. Radio waves are refracted in the middle and upper corona. The result is that the observed solar diameter, which is 32 minutes of arc on the average ($\approx 0.5°$) in the visible, reaches 0.75° at 600 MHz (50 cm of wavelength for approximately 7 400 Km above the photosphere) and 1° at 60 MHz (500 cm of wavelength for approximately 210 000 Km above the photosphere).

Professionals have turned this difficulty by using the **interferometry** principle. This process consists in using two antennae located at a distance B of each other. Signals are mixed and when they arrive in phase on the mixing device, their respective amplitude is added, and subtracted when they arrive in opposition of phase (the signal received on each antenna at the same instant is identical in amplitude, but not in phase because the Sun is moving). This process allows to get interference fringes (alternation of darker and brighter zones) whose distance between each of them (the interfrange) is proportional to the wavelength and conversely proportional to the basis B. The more the basis is extended, the more the interfrange is easily measurable; but correlatively, the width of the fringe is more and more narrow, therefore less and less contrasted, hence less and less detectable above the noise level (in practice, an optimum is explored). One can add several other antennae to research interference fringes two by two. The calculation, by Fourier transform, shows that it is then possible to reconstruct the whole image of the source. The image shown in Figure 2 was obtained on this principle. Today, mathematics details are accessible to qualified amateurs.

One can without no doubt consider as regrettable this lack of amateur enthusiasm for the radio wavelength range[4], and more especially as the spectrum reveals at least two surprises:

• The first one points out the 10.7-cm flux which is a very good indicator of the solar activity cycle. By contrast to other classic indexes, calculated according to the number of spots and faculae observed on the surface of the Sun (formerly called Wolf number, or International Sunspot Numbers[5]), which is only a number (even if it represents a fraction area of the "obscured" Sun), the radio flux (measured in Ottawa since 1950) is a true physical quantity. One will put this index in close relation with that drawn from the observations of the two Magnesium lines at 278.6 nm and 280.1 nm: see Figure 9.

• The second issue comes from the emission of what is called the "active Sun" (by opposition to the quiet Sun). Electrons (and sometimes protons and other ionized particles) are accelerated by the solar magnetic field, describing spirals around the lines of force. These accelerated electrons emit a radio radiation called **synchrotron** radiation. During eruptive manifestations of the Sun, it may occur that electrons reach a speed close to that of the light, and are thus violently ejected from the corona to the outer space. Such radio emissions take the name of **"storms"** and their type is codified according to the manner whose frequency

---

[4]For an equipment of quality, one may usefully consult the site: `//www.fm-transmitter.com /kits/amateur-radio/aerials/140-300-MHZ-YAGI-AERIAL.htm`.

[5]See for example the solar database at `http://www.sidc.be/products/qua`.

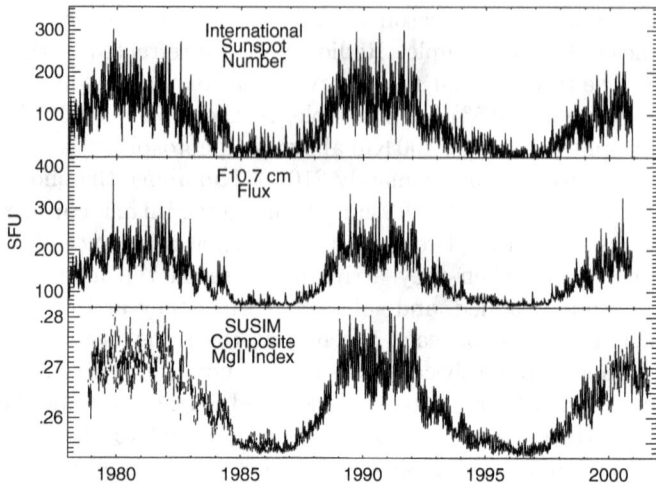

Fig. 9. Three indexes of the solar activity cycle. Top, the classical indicator, a ponderated function of the number of spots and faculae observed on the Sun, a useful proxy, but without true physical significance. Center, the radio flux at 10.7 cm. Down, the relationship between the central flux and the flux in the wing of the Mg I and Mg II lines, close to 280 nm. This last proxy, which has a physical meaning, appears to be very useful for modeling the solar activity prediction. Fortunately, these three indexes are highly correlated. (According to Floyd 2003.)

varies with time. These storms are often the forerunner sign of violently disruptive phenomenon in the corona (called **"flares"**) and they can reach –and release– considerable energies. Radio storms travel into the interplanetary space and may interact with the terrestrial atmosphere, giving **boreal auroras**. The observation of radio storms is very important today because they can be considered as a confident indicator in predicting phenomena playing a role in space weather.

The tools of radio astronomy nicely complement the solar X-ray and gamma-ray emissions views. This is because radio waves are emitted by energetic electrons of comparable energies, and because radio telescopes can be extremely sensitive. The radio spectrum is in fact vast –the accessible wavelength range being something like 1 mm to 10 km– and each different band tells us something about a different part of the solar atmosphere. "Decimetric waves", *i.e.* 0.1 m wavelength (or 1 GHz frequency) are more often studied, because both hard X-ray and decimetric radiation are emitted during flares, and both are widely believed to originate from non-thermal electrons. One would expect that the two emissions correlate well with each other. They do not in general and contrary to the case of solar centimeter emissions, caused by incoherent gyro-synchrotron emission, the decimeter waves, are emitted by coherent processes. Here "coherent" means that the emitting electrons have correlated motions, which can make their emissivity much larger than that of the same number of independently moving ("incoherent") electrons.

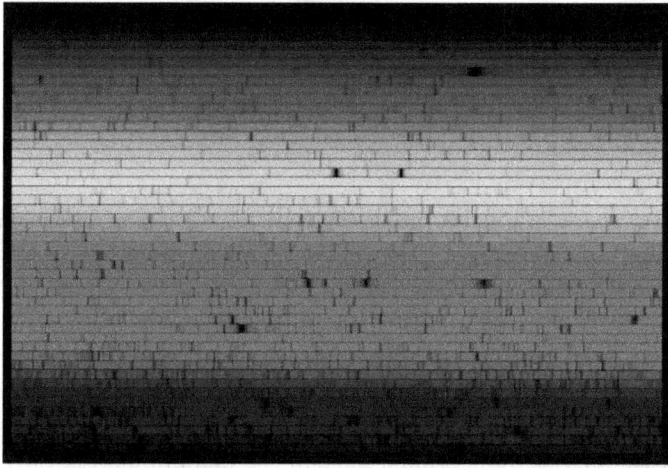

Fig. 10. Spectrum of the Sun in the visible field. Black (vertical) lines are called Fraunhofer lines and are specific of the emissive element.

Only occasionally do some of these coherent decimetric emissions coincide with the hard X-rays. Some good correlations have been reported in the past, but it appears today that the radio emissions and hard X-rays often originate from different populations of accelerated electrons (Dabrowski & Benz 2008).

2.2.2   The visible field

The study of the visible spectrum is rich as it can be noticed on Figure 10. We will put here apart the study of the chromospheric and coronal spectra, which is another rich topic, as well as the aspects linked to the spectrograph principles, a matter approached by C. Buil in this volume.

All amateur desirous to tackle the solar spectrography may begin by building for example a Czerny-Turner type spectrograph (Fig. 11), whose easiness to assembly is equal to its compactness and its good luminosity. One can note that the solar tower Mac Maths-Pierce of Kitt Peak (USA) spectrograph is of this type. One can see also the MERIS or LHIRES spectrograph types (Chapter by C. Buil).

In the focal plan of the telescope (equipped with an attenuator filter!), the image of the Sun is formed on the entrance split of the spectrograph. The light cone is shaped parallel by means of a collimator, and falls on the grating. The dispersed light is re-focused on the exit slit by a second mirror located below the collimator. The whole system is very compact and, to our opinion, can be easily used to record spectroheliograms. Indeed, it is sufficient to turn gradually the package around the center of the entrance slit to obtain a quasi continuous image of the observed spectrum. The Czerny-Turner mounting type is very well adapted to this type of operation, which may give great satisfaction to his owner. Images of the Sun obtained in a peculiar wavelength, then shifted by + or − some nm, lead

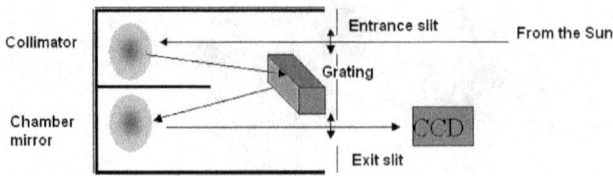

Fig. 11. Spectrograph of Czerny-Turner type, one of the easiest to manufacture. All adjustments are easily accessible. That is the case for example for the three "pulls-shoot" screws holding the mirrors (to the rear), for the adjustment toothed wheel of the grating and for the screws which permits the opening (or closing) of the entrance and exit slits. Moreover, it can be easily balanced, because very symmetrical. From an optical point of view, the collimating beams insures a good luminosity.

with some basic physics to determine some of the main properties of the observed medium. An image of that type is given in this book (chapter by C. Buil) and can be also found in Rondi *et al.* (2006). Properties and aberrations of this type of spectrograph can be found in Shroder (1987; see pages 276 to 281)[6].

All solar spectral lines can be thus analyzed. And they are numerous! (see Fig. 10). Methods of analysis are the same that for any star spectral lines (see the chapter written by A. Acker in this volume). One of the results, always sensational to obtain, is the temperature of the medium in which the line is emitted. As an example, let $\delta\lambda$ the half width of a line centered at wavelength λ, then:

$$T = 1.95\ 10^{12} \times M \times \left(\frac{\delta\lambda}{\lambda}\right)^2 \qquad (2)$$

where M is the molecular mass of the ion in question. In the case of the green line excited in the coronal medium, $\lambda = 530.3\,\mathrm{nm}$ and the measurements give $\delta\lambda = 0.08\,\mathrm{nm}$ as an average. This line being emitted by the Fe XIV ion, for which $M = 56$, it ensues that:

$$T = 2.48 \quad \text{millions of } °\mathrm{K} \quad (\text{Kelvin}).$$

2.3   The Zeeman effect

The presence of the magnetic field alters the spectral lines by splitting them into several components around the initial wavelength, according to an effect found by Zeeman in 1896[7]. By measuring the interval of the splittings, it is possible to deduce the intensity of the magnetic field. George Hale applied for the first time this physical effect on lines recorded in a sunspot spectrum. Thus, he has been

---

[6]Solar spectrographs are available at Shelyak Instruments for instance.

[7]Pieter Zeeman (1865–1943), Dutch physicist who shared the 1902 Nobel Prize in Physics with Hendrik Lorentz, for this discovery.

able to show that spots hold a strong magnetic field, and in any case, stronger than the surrounding photospheric one.

When an atom is dive into a magnetic field $\vec{B}$, it behaves as a compass, because all the electrons contribute to give a total magnetic moment $\vec{M}$. If this "magnetic compass" is aligned on the direction of the field $\vec{B}$, the energy of electrons increases, but decreases in the opposite case. However a variation of energy of an electron corresponds to a change of the emitted wavelength. A spectral line emitted by a group of atoms sees a global change which will be characterized by a change of state of the departure level, so this latter one split. As compared to the initial state, there are three possibilities for the electron to return from these new levels to the fundamental. The line is therefore seen under the form of a *triplet* of lines, obviously of very close wavelengths. Each of the two levels excited in the presence of the magnetic field is very close to the non excited level, so that the difference of wavelengths is small. If the magnetic field is not enough strong, levels are only weakly split and the effect is not perceptible. To first approximation, the distance that separates the components is proportional to the intensity of the magnetic field. One has therefore there an efficient and simple way to know the intensity of the solar magnetic field and this property is finalized in instruments such as solar magnetographs.

Moreover, the emitted light has a **polarisation** which depends on the magnetic field. If this one is longitudinal, and directed to the exterior (parallel to the line of sight, *i.e.* to yourself), the shifted line has a circular right polarisation. If the magnetic filed is longitudinal, and directed to the interior, the shifted line has a circular left polarisation. Finally, if it is transverse (perpendicular to the line of sight), the polarisation is linear. With the help of polarised filters, it is therefore possible to know the direction of the magnetic field.

For people who are interested in quantum physics (see also chapter written by A. Klotz), the wavelength difference λ, as compared to the wavelength λ_0 of the "non magnetic" state is given by:

$$\lambda - \lambda_0 = \frac{e}{4\pi c m_e} g^* \lambda_0^2 B \qquad (3)$$

where $g^* = gM - g'M'$,

is the Landé factor[8] of the transition when going from the non excited state $\vec{M}$ to the excited state $\vec{M'}$. This Landé factor can be written quite simply for each state, as a function of the numbers L, S and J characterizing the levels of energy. L is the global orbital angular moment, S is the total spin and J is the total angular moment. M is the quantic magnetic number, which can take only the values

---

[8]The Landé factor is given by:

$$g = 1 + \frac{J(J+1) + S(S+1) - L(L+1)}{2J(J+1)}$$

and tables have been compiled for each atom. See for example Beckers (1969).

$-J, -J+1, \ldots, J$, while J itself can take only the values $|L - S|$, $|L - S| + 1, \ldots,$ $L + S$. If the magnetic field $\vec{B}$ is null, then all the M are null. This degeneration is raised in the opposite case, that gives a shift in energy of the excited level.

In the simple case where the fundamental and the excited state are characterized by $S = 0$, then, by virtue of the selection rules stated higher, $\Delta M = -1$, 0, 1, that leads to $g^* = -1$, 0, 1. It is the "normal" Zeeman effect (one says sometimes Lorentz triplet). This feature remains even so enough general, so that the Zeeman triplet consists of two components σ, representing the shift levels on each side of the non perturbed state, which takes the name of π component. In the other cases, only multiplets can be seen.

If λ is measured in cm and $\vec{B}$ in Gauss, then

$$\lambda - \lambda_0 = 4.67 \ 10^{-5} g^* \lambda_0^2 B. \tag{4}$$

Fraunhofer solar spectral lines being already widened by Doppler effect[9], they must have a magnetic field $\vec{B}$ minimum, which can be set up at around 150 Gauss (0.15 T). Georges Hale had in fact a lot of chance with his discovery in 1908, because the measured field was 3 300 Gauss (3.3 T). In reality, the magnetic field is very often at the detection limit, as indicated by the following example: Fe line for which $\lambda = 525.022$ nm; $g^* = 3$; measured $\Delta\lambda = 42$ Å. The formula 4 gives:

$$\vec{B} = 110 \ \text{Gauss} \ (0.11 \ \text{T})$$

Attention! The rules giving the direction of the polarisation in the case of a normal Zeeman effect (case of a triplet) are valid for lines in *absorption* (native from an optically thin plasma). For lines in *emission*, the direction of circular polarisation is inverted (perpendicular and parallel terms have to be exchanged). The reason is that the emission lines have an intensity by themselves while what is seen in absorption is a residual intensity.

3   The "resonant" spectrum

Before the sixties years, the idea would not have come to any one that it would be possible, from the ground, to probe the interior of the Sun. A bit fortuitously (but not completely), at the eve of these years, Robert B. Leighton, Robert W. Walnuts and Georges W. Simon, were observing the photosphere by Doppler techniques (see footnote 9). They observed unexpected vertical movements in the medium, which did not vary in time on a random way, but showing a periodicity of 5 minutes, and to be more precise, of 296 ± 3 seconds. These oscillations were

---

[9]The Doppler effect is the shift in wavelength of a wave as perceived by an observer when the source is moving with respect to the medium. A red shift occurs when the source and observer are moving away from each other (recession speed), and a blue shift occurs when the source and observer are moving towards each other. The red shift of light from remote galaxies proofs that the universe is expanding.

localized over zones of some thousand of kilometers. The speed was 500 meters per second. Each regions seemed to oscillate independently of each other. This phenomenon was attributed to resonant waves created and ordered to small scale motions, of ascending and descending structures of hot gas localized in convective cells called **granules**. These granules, of oblong form, of size of approximately 1000 Km, had been the object of intense research in the 1960 years, especially at the Pic du Midi Observatory (by means of the 50-cm refractor set up inside the so-called Turret dome), especially by Jean Rösch and Marcel Hugon and later on by Richard Müller: thanks to images of exceptional quality, they have been able to attend the "birth" of granules, under the form of what they have called "pores". But they went over the discovery of the 5 minutes oscillations, of which neverthe-less everyone was speaking a lot during lunch's time... It has been necessary to wait the years 1970–1971 to understand that such oscillations found originally at 5 minutes (and later on to others values, 160 minutes for example), had nothing to see with the granules themselves, but were resonant modes, therefore (more) stable, of several thousands of others waves, in fact more than 10^7, taking birth inside the Sun itself.

Roger Ulrich, from the University of California in Los Angeles (USA), has been the first to formulate the quasi complete theory in 1971. Since then, *heliosismol-ogy* was born, which will give to the study of the Sun a vigorous new impetus. Results were rapidly transferred to other stars, and new theories thus developed are known now under the vocable of *asterosismology*.

Thousands of **acoustic waves** propagate inside the Sun. Each of them pos-sesses its own wavelength and its own propagation path. But the wave does not move straight on, because the sound speed increases with depth, and thus the wave is gradually refracted[10]. At a depth r_c, the wave is completely refracted and turns back. When arriving at the surface, the brusque diminution of density will force the wave to be reflected to the interior (beyond, it is the "empty space" and the sound does not propagate in the vacuum). The wave is thus trapped in a spherical cavity delimited by the surface and the sphere of radius r_c. Then, it can turn endless between the two inner shells (this is called a resonant cavity). But it may happen also that the wave has just the adequate length to describe an integer number of rebounds around the solar circumference.

For example, the 5 minutes oscillations result from the superposition of hun-dreds of individual vibrations: each of them displace the solar surface vertically, of a few meters only, the speeds being of some centimeters per second. But if these waves are phased, their interferences intensify the phenomenon, up to around 500 meters per second, and canceled it if they are in phase opposition. In the first case, they become detectable from the ground.

Finally, it remains to understand why these acoustic waves are excited and are temporal persistent (why they are not damped since a lot of years?). The excitation

---

[10] The sound speed c is $\propto T^{1/2}$, and the temperature decreases from the center of the Sun to the surface; thus the propagation is not rectilinear.

comes from the violently turbulent nature of the gas inside the convective zone of the Sun: it is the same phenomenon which render noisy the agitated and hot gases at the exit of a plane reactor... The persistence is due to the fact that the solar nuclear core maintains the heating. Indeed, the seismologic analysis will reveal two surprises, one lying between the frontier of the radiative core and the convective zone, and the other just below the surface (see paragraph describing the tachocline and the leptocline).

3.1 Modes distribution

Under the solar surface, a propagating acoustic wave describes a regular festoon and is characterized by a degree l which is equal to the number of inscribed arcs; the speed of the wave is canceled at each reflection point on the surface. The greater l is, the less deeply the wave penetrates, and reciprocally. Waves which remain close to the surface are **pressure waves**, called p-modes, because the governing restoring force is essentially of a typical Archimedes force. Waves that originate in depth, called g-modes, are **gravity waves**, because the governing force is essentially of gravitational type.

The depth of penetration d (which defines the sphere of radius r_c) varies approximately as

$$d = (2n + 3)/k_h$$

where n is an integer number known as the **radial order** of the mode and k_h is the **horizontal wave number**. This horizontal wave number k_h is connected to l by:

$$k_h = \frac{[l(l+1)]^{1/2}}{R_\odot} \tag{5}$$

where $R_\odot$ is the radius of the Sun. Finally, for each value of l, one can associate an **azimuthal order** m taking the values $2l + 1$. The horizontal wave number defines the size of the oscillation, that is to say the region of the Sun where the oscillation will be visible. These regions are animated like palpitation motions, sometimes appearing to advance to the observer, sometimes retreating, of a few meters. Nowadays, oscillations of a few millimeters can be investigated, which permit to get information on the rotation speed of the solar core.

Each mode p or g, is therefore characterized by three numbers, l, n and m. As far as these values are increasing, the zones on the Sun where the pressure is exerted are more and more complex. Figure 12 shows such examples.

Just before ending this paragraph, and for people that mathematics do not rebuff too much (without going into further details, because it would be necessary to use more complex tools such as spherical harmonics, which better describe this subject), we may say some words to understand the distinction between p-modes and g-modes. In spherical symmetry, the general propagation equation of a wave Ψ is written as:

$$\frac{d^2\Psi}{dr^2} + \frac{1}{c^2}[Z]\,\Psi = 0, \tag{6}$$

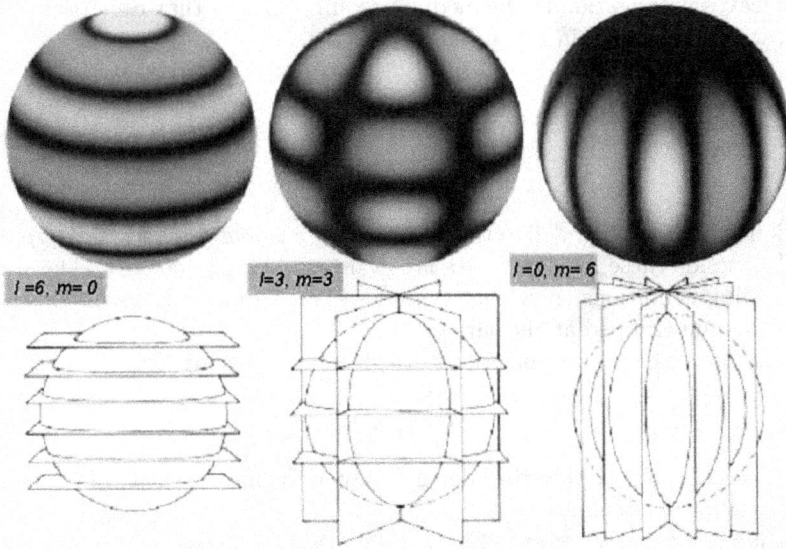

Fig. 12. Example of solar oscillations. On the left, $l = 6$ and $m = 0$ indicate that the points where the speed of the wave cancel, are distributed on 6 different latitudes. On the right, $l = 0$ and $m = 6$, exhibits on the contrary, the points where the waves are distributed along 6 meridians. Center, $l = 3$ and $m = 3$ show an intermediate system.

where c is the sound speed in adiabatic conditions (solar case) and Z depends on the conditions of the medium. In the case of the Sun, it can be shown that:

$$Z = \left[\omega^2 - \omega_c^2 - S_l^2 \left(1 - \frac{N^2}{\omega^2} \right) \right],\tag{7}$$

where ω is the frequency of the studied wave, ω_c^2 is the cutting frequency, directly linked to the density of the medium[11], S_l is the Lamb frequency connected to Equation (5) by:

$$S_l^2 = \frac{l(l+1)c^2}{R_\odot^2},\tag{9}$$

and N is the frequency of the back force[12].

---

[11]The cutting acoustic frequency ω_c is defined by:

$$\omega_c^2 = \frac{c^2}{4H^2} \left(1 - 2\frac{dH}{dr} \right),\tag{8}$$

where $H = -(d \ln \varrho/dr)^{-1}$ is the density scale height.

[12]This frequency is defined by:

$$N^2 = g \left(\frac{1}{\Gamma_1 p} \frac{dp}{dr} - \frac{1}{\varrho} \frac{d\varrho}{dr} \right),\tag{10}$$

The waves propagation in the medium requires clearly that Equation (7) must be > 0, which is verified in two fields described as:

$$\omega^2 > S_l^2 \quad \omega^2 > \omega_c^2 \tag{12}$$

and

$$\omega^2 < N^2. \tag{13}$$

Conditions (12) and (13) define the *regions where p and g modes are respectively trapped.* Outside these regions, waves are evanescent or do not exist. The study of the preceding conditions show that *g*-modes allow to probe the solar core, while *p*-modes remain trapped at the surface.

Finally, it is easy to see that the ω frequencies correspond to:

$$\omega^2 \simeq g_\odot k_h, \tag{14}$$

where $g_\odot = GM/R_\odot^3$, so that they depend only of the mean density of the medium, and not of its thin structure.

To conclude this section, the Sun can be also considered as "a laboratory for quantum physics". With the advent of highly sensitive imaging devices such as polarimeters, magnetographs and spectrographs, an entirely new "spectral face" of the Sun has become accessible to exploration. It is due to coherent scattering processes, which produce a spectrum that is as richly structured as the ordinary intensity spectrum but with spectral structures that look entirely different and that have different physical origins. The work on trying to identify these various, previously unfamiliar structures has led to new insights in atomic and quantum physics. Spectral signatures of various types of quantum interferences, hyperfine structure, and optical pumping has been found. The molecular lines, which are very weak and next to invisible in the intensity spectrum, stand out with high contrast in the so-called "Second Solar Spectrum". There are also structures that have remained enigmatic for more than a decade, an example being the observed polarization peak in the D1 line of sodium. According to quantum mechanical predictions this line should be intrinsically unpolarizable. To determine whether this is a problem of solar physics or of quantum physics, experiments has been set up to explore the properties of the polarized D1 scattering under controlled conditions and in well defined magnetic fields. This experiment has produced unexpected results that are unequivocally at odds with our current understanding of quantum scattering. The second solar spectrum has also given a diagnostic tool that allows to explore aspects of solar magnetism that have been inaccessible

---

Γ_1 being the first adiabatic exponent of the medium (partial logarithmic derivative of the pressure as compared to the density at specific constant entropy:

$$\Gamma_1 = \left(\frac{\partial \ln p}{\partial \ln \varrho} \right)_{ad}. \tag{11}$$

The solar plasma being a quasi perfect gas, Γ_1 is close to 5/3 in the quasi totality of the solar interior.

to the Zeeman effect. Thereby vast amounts of "hidden" magnetic flux in the photosphere is been to be uncovered, and leads to a new view of the nature of solar magnetism.

3.2 Measurements benefit

The detection of p and g modes gives access to well defined regions of the solar interior. Measurements are now made regularly from the ground and from space. The usual way to observe these modes and to detect global resonances is made by recording the shift through the Doppler effect, of a specific absorption line, in integrated light on the disc, over a very long period of time, in order to suppress crests and depressions of random oscillations that occur in the observation field. For example, 8 hours of observation allow to solve spectral lines of 34.72 μHz; with 30 days, one reach 0.3858 μHz and in 1 year, 0.03169 μHz. The difficulty comes from the days and night alternation; 24 hours give an oscillation of 11.57 μHz, that obviously is not of solar origin. This is why measurements were also made to the South pole (by E. Fossat and G. Grec –from the University of Nice, F– and M. Pomerantz– from the Delaware, US–), during 5 uninterrupted and consecutive days, in 1983. Results shown that the Sun can vibrate as a bell during entire days...

A different approach has been used by G. Isaak, from the University of Birmingham (UK), by setting measurement apparatus spaced in longitude, such as Tenerife, Hawai, Pic du Midi. This allows to cancel the day-night alternation, by "sticking" measurements in time. Nowadays, such systems work permanently, allowing a large collection of results. For the observations, the most frequently line used is that of the Sodium, Na D1 at 589.6 nm. A change of the wavelength $\Delta\lambda$, of 2.10^{-5} nm, allows a solar oscillation measurement of 1 m/s. Others lines can be used, for example that of the Potassium, at 769.9 nm which gives also a good photometric signal. These instruments, based on the lines shift, have a very good temporal stability. On such measurement questions, the reader may consult the paper written by Bernard Gelly (2003).

The analysis of the observed "resonant spectra" is not simple. Two important results have been highlighted:

• The first one points out the *rotation velocity rate* of our star. The so-called data inversion, allows to estimate the rotation velocity rate, at the surface and in depth. Figures 13 and 14 shows that the Sun is far from rotating uniformly. The important discovery is that the differential rotation velocity, found from the first measurements made by Galileo, by means of sunspots observation, goes on in depth, at least until 0.7 $R_\odot$. It is likely that the solar magnetism begins to act from there, according to physical processes that one begins hardly to understand. Let us note the long way covered since the observations made in 1635 shown on Figure 16! For further details, especially on the data inversion process, one may consult the paper written by Maria Pia di Mauro (2003).

• The second issue is the existence of the **tachocline**. The difference between the measured sound speed and its theoretical value, plotted as a function of the

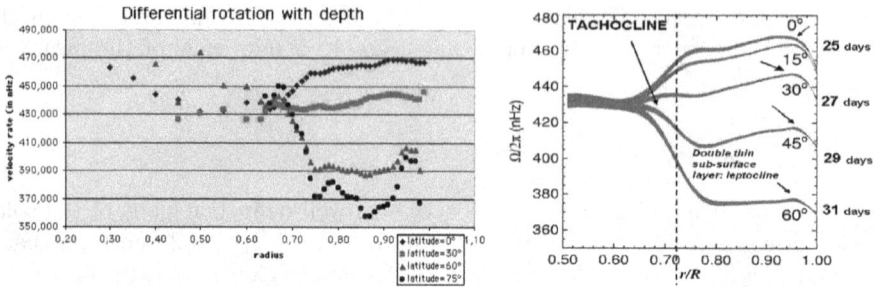

Fig. 13. *Left*: first differential measurements of the rotation deduced from heliosismology. It appeared very clearly that the differential surface rotation is anchored in depth, at least until to the level of the tachocline (at 0.7 $R_\odot$), which is now considered as the seat of solar magnetic effects (After Maria Pia Di Mauro 2003). *Right*: modern measurements made just at the surface, where the error bars were great at the onset of the observations, and being reduced due to the longer time of the observations. See Howe (2000) and Rozelot *et al.* (2007). The arrows show a break at 0.99 fractional radius, which is the signature of the leptocline, cradle of solar asphericities, radius variations with the 11-yr cycle and complex physical processes: partial ionisation of the light elements, opacities changes, superadiabaticity, strong gradient of rotation and change in the turbulence pressure.

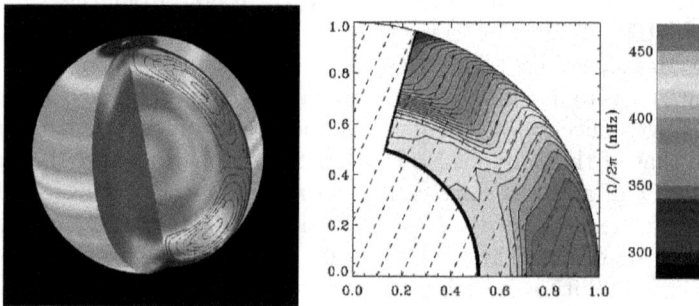

Fig. 14. *Left*: 3-D view of the solar differential rotation developed at the surface and acting in depth until approximately 0.7 $R_\odot$. *Right*: 2-D view showing the different velocity rates. It is now acknowledged that the solar core may have a solid rotation of about twice that of the surface.

distance from the surface to the center of the star, shows a "bump" at 0.7 $R_\odot$. This region, extremely thin (of approximately 0.05 $R_\odot$, *i.e.* $\approx 35\,000$ km) makes the junction between the interior of the Sun, radiative, rotating at a uniform speed (and likely more rapidly than at the surface), and the most external layer, convective, rotating at different speeds. Just below the surface, a new very thin layer has been recently put in evidence. This sub-surface shell, called the **leptocline** (Fig. 15), insures a new passage from a convective zone to a radiative one, just before the photons escape in space. One can show, by computing the solar gravitational moments (directly linked to the moments of inertia), that the envelope of

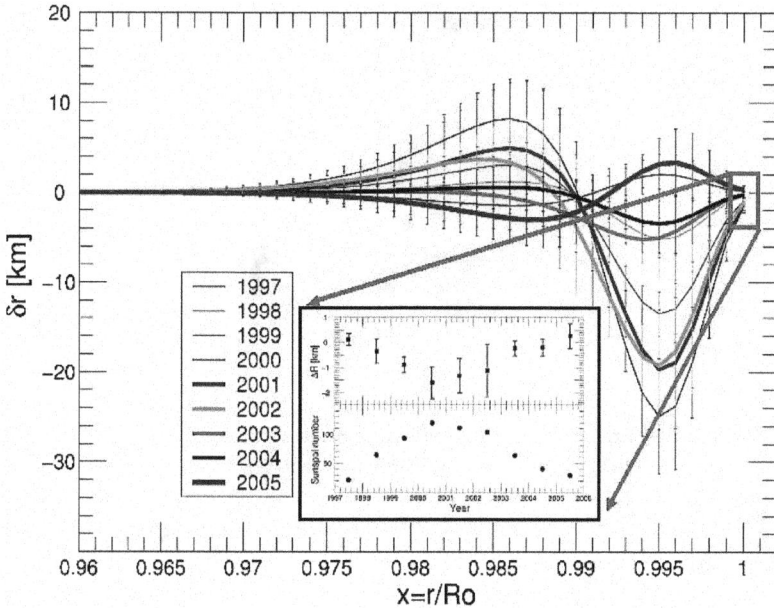

Fig. 15. Radial variation δr as a function of the fractional radius $x = r/R_\odot$, obtained as a solution of the inversion of f-modes (see paragraph 3.1 and Eq. (14)). The box shows a zoom of the variation at the surface, compared with the sunspot number. The reference year is 1996. After Lefebvre *et al.* (2007). One can see: (i) A non-homologous variations of the position of the subsurface layers with depth; (ii) a change in position going from being in phase with the solar cycle in the deeper layers to antiphase in the shallower layers with a transition at $0.99\ R_\odot$ and a maximum of contraction during the peak of activity at $0.995\ R_\odot$. The variation of the radius just at the surface is found to be in antiphase with the solar cycle, with an amplitude of about 2 km.

the free surface is distorted: a small equatorial bulge is followed by a small depression, just above the *royal zones* (*i.e.* zones lying between around $-45°$ and $+45°$ of heliographic latitudes, where sunspots born, live and disappear); the whole envelope presents a spheroidal shape. Figure 17 shows the last progresses concerning the shape of the Sun. Departures to pure sphericity do not exceed some 20 mas as a maximum estimate. Even if this order of magnitude is faint, it remains of astrophysical interest to measure the solar diameter in all heliographic latitudes (see Sect. 3). For instance, such small departures may have repercussions on the general relativity (Pireaux & Rozelot 2003), but this would take us here too far...

In conclusion, if the study of the electromagnetic spectrum allows us to probe the solar atmosphere, the study of the resonant spectrum, in the whole range of frequencies, permits to probe the totality of the interior of the star. As far as the author of this paper is concerned, it is precisely the interface between these two regions (outer and inner) which is of scientific interest today...

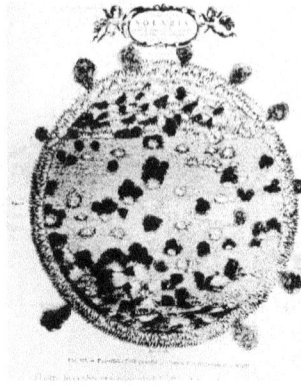

Fig. 16. Facsimile of an engraving representing sunspots during the year 1635 by Scheiner (1575–1650, a Jesuit father). Since a long time ago, sunspots have been seen several times to the naked eye, in China and in Europe. Virgil (classical Roman poet), for example, states in the *Georgics* (written 37–29 BC): "Sin maculae incipient rutilo immisceri igni". Such a large group of spots was also visible to the naked eye on Saturday, 3 May 2003.

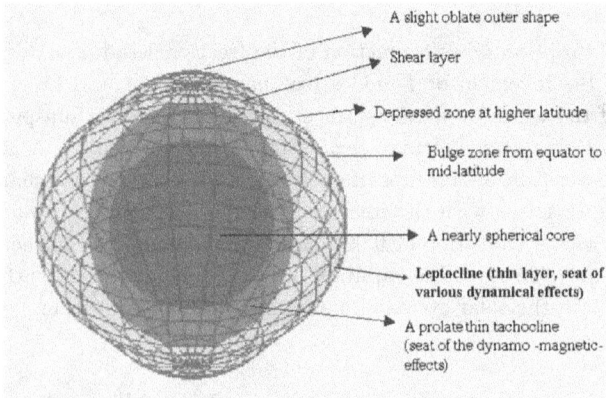

Fig. 17. Shape of the Sun, according to Rozelot & Lefebvre (2003). Sizes are strongly exaggerated. One can see a faint equatorial bulge, followed above the royal zones at around ±45–50°, by a slight depression, the whole Sun remaining as a spheroid (*i.e.* the shape, close to those of an ellipsoid, that is to say the figure taken by a fluid rotating at a uniform speed). Departures to the sphericity are very faint, not exceeding some kilometers (see Hudson & Rozelot 2010).

4 Solar shape determination applied to stars

In the same way that the "geoid" is defined for the Earth, which is an equipotential surface of gravity, we can define the "helioid" for the Sun and a "stelioid" for the other stars. These words illustrate the fact that the surface of the Sun and stars is distorted under their own rotation and possibly by their magnetism. For the Earth, precise space measurements of the surface shape allow us to get information on the internal terrestrial structure, mainly the density inhomogeneities or the angular momentum variations. The idea to apply the same method to the Sun and now, to the stars is on the way. In such cases, the study of the outer shape of the body permits to access not only to density inhomogeneities under the surface, but also to the different rotation velocity rates, which are known to be non uniform, both at the surface and in depth. With the advent of sophisticated techniques such as interferometry, it is now possible to accurately determine the geometrical shape of the free boundary of stars, such as Altair or Achernar, two rapid rotators for which observations of the geometrical envelope have been made by Belle *et al.* (2001) and Domiciano *et al.* (2003) respectively. Curiously, for the Sun, the situation is hardly that of the Earth which prevailed before 1957. The flattening of the Sun is still poorly estimated and we hardly know if there is a variability of this flattening related to the solar cycle[13]. One of the main reasons is that the Sun is a slow rotator. If there are no surface stresses, such as those introduced by the fluid velocity rates distribution or magnetic fields, then the Von Zeipel's theorem states that the gravitational surface coincides with those of constant pressure or constant temperature. The visual (or apparent) figure is a surface of constant gravitational potential, which shows small departures from a perfect sphere, called **asphericities**, and this equipotential level must match the physical Sun's surface. If we are able to determine by accurate observations such asphericities, then we will learn about the solar interior by adjusting the perturbated gravitational field (a fact that G. Isaak describes as *"a new window open over the Sun's interior"*, see Godier & Rozelot (2001); such a program is part of the objectives of new space missions). In many cases, it is sufficient to model the Sun as a sphere to explain most of the physics. But we are now arriving at such a level of precision, that it becomes necessarily to go further, particularly in describing the subsurface layers (see the paragraph on the leptocline), where the behavior of the Sun is still questioning (mainly due to density variations and to the change of the sign of the rotation rate $d\Omega/dr$ at mid heliographic latitudes: $d\Omega/dr$) is <0 from 0° to 45° and >0 after). This is why, solar and stellar models need accurate measurements as a function of the colatitude θ of the following parameters:

- the radius, $R(\theta)$, (*i.e.* the precise shape of the studied bodies)

- the effective temperature, $T_{eff}(\theta)$ and

- the rotation law, $\Omega(r, d, \theta)$, where d is the depth.

Presently, advanced techniques such as interferometry lead to a precision never achieved for the oblateness of (rapid) rotator stars. For example McAlister *et al.* (2005) have recently determined the oblateness of Regulus (αLeo) to be

---

[13]Through recent studies, it is now proved that the oblateness follows the solar cycle (first mentioned by Rösch *et al.* 1996, then by Emilio *et al.* 2007 and by Rozelot & Damiani 2009).

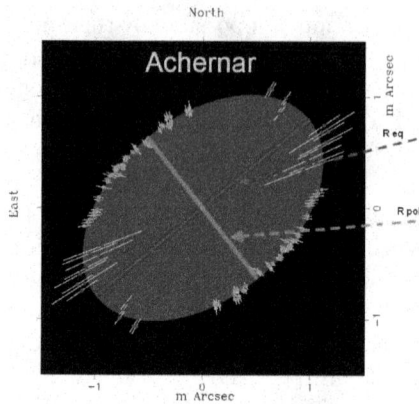

Fig. 18. The oblateness of α Eri (Achernar) as determined through interferometry at Paranal (ESO observatory) by A. Domiciano de Souza *et al.* (Astron. & Astrophys., 407, L47). The difference between the major and minor semi axis is 910 ± 50 microas (*i.e.* 5.5%).

$a/b = 0.845 \pm 0.029$, where a is the equatorial radius and b the polar one. Domiciano *et al.* (2003) were the first to measure the shape of Achernar (αEri) (Fig. 18), the flattest star ever seen: the best ellipse fits leads to $a/b = 1.56 \pm 0.05$ or $2a - 2b = 0.91 \pm 0.05$ mas which gives a relative precision of 5%! Belle *et al.* (2001) have measured the size of Altair (αAqu) and obtained a value $a/b = 1.140 \pm 0.029$ or $2a - 2b = 0.424 \pm 0.079$ mas (19%) (see Table 1). For the Sun, Rozelot *et al.* (2003) obtained accurate values of the solar shape by means of the Pic du Midi heliometer: they measured the difference $a - b$ to be 9.45 ± 1.41 mas (as an average), that is 15% of relative precision, less than the precision for Achernar (other measurements exist, see a review in Godier & Rozelot (2000). Thus, we are presently confronted to a paradox: a better determination of the shape of other stars than that for our own Sun exist (see also, Domiciano de Souza 2008; Royer 2008).

We are convinced that the amateur can be able to determine today with a great accuracy the radius of the Sun in any heliographic positions, and thus may help to significantly improve our knowledge in this field.

5 Observing the Sun

5.1 A fundamental quantity: The solar diameter

Since the highest antiquity, men have searched to determine the size of the universe, the size of the Earth, those of the Moon and in particular those of the Sun. For the Sun's size, they achieved it with a relative success, since Aristachus of Samos

Table 1. Interferometric measurements of different stars diameters. Major axis: 2a. Minor axis: 2b.

- Rattenbury *et al.* (2005): MOA-33

a/b = 1.02 +0.04, −0.02

- McAllister *et al.* (2005): Regulus (α Leo)

a = 0.771; b = 0.651; b/a = 0.845 ± 0.029

- Domiciano de Souza *et al.* (2003): Achernar (α Eri)

2a = 2.53 ± 0.06 mas *i.e.* R_{eq} = 12.0 ± 0.4 $R_\odot$(= 8.4 M km)

2b = 1.62 ± 0.01 mas *i.e.* R_{pol} = 7.7 ± 0.2 $R_\odot$(= 5.4 M km)

2a − 2b = 0.91 ± 0.05 mas (5%)

- Belle *et al.* (2001): Altair (α Aqu)

2a = 3.461 ± 0.038 mas; 2b = 3.037 ± 0.069 mas

2a − 2b = 0.424 ± 0.079 mas (19%); a/b = 1.140 ± 0.029

- Richichi & Roccatagliata (2005): Aldebaran (α Tau)

Diameter = 20.58 ± 0.03 mas (limb-darkening diameter)

- Di Folco *et al.* (2004): Vega-like stars

τ Cet: R = 0.816 ± 0.013 $R_\odot$

ε Eri: R = 0.743 ± 0.010 $R_\odot$

β Leo: R = 1.728 ± 0.037 $R_\odot$

β Pic: R = 1.759 ± 0.241 $R_\odot$

- Chesneau *et al.* (2005): Be stars

α Ara: R = 4.8 $R_\odot$ *i.e.* an envelope size of 4.0 ± 1.5 mas

- ESO press release November 5th 2001:

Red dwarf HD 217987: Diameter = 0.92 ± 0.05 mas

Giant star HD 36167: Diameter = 3.32 ± 0.02 mas

Cepheid ζ Gem: Diameter = 1.78 ± 0.02 mas (variations between 1.5 and 1.8 mas in 10 days)

Cepheid β Dor: Diameter = 2.00 ± 0.04 mas

Red giant ψ Pho: Diameter = 8.21 ± 0.02 mas

(circa 310-circa 230 BC[14]) evaluated the diameter of the Sun to be 1/270 fraction of the right angle (that is to say 0.45°), while Archimedes (287–212 BC) gave a value ranging from 1/ 200 to 1/ 154 fractions of this same right angle (that is to say between 0.45° and 0.58°). These measures enclose the currently admitted value (0.53°)! Paradoxically, today, the absolute value of the solar diameter is not yet accurately known, by comparison with the accuracy that modern techniques may achieved. The most recent ephemerides book (Allen 2000), gives $R_\odot$ = 6.955 080 ± 0.000 26 10^8 m. The Sun is thus thinner of some 700 km when comparing this value to those given in the 1980 ephemerides and before. The reason of this cure of thinness comes from helioseismology, as the seismic diameter lies under the photospheric one (see paragraph 3). For this last parameter, the amateur will keep

---

[14]BC: Before our era, *i.e.* the Christ.

in mind the value of $s=$ 959.63 seconds arc. This unit is still (and unfortunately) used by astronomers, indicating that it is the angle under which one sees the solar diameter at a distance of 1 Astronomical Unit (UA). This is true, as if one takes
$d = 1\,\text{UA} = 1.495\,978\,706\,610^{11}$ m, then:
R_{app} (in km) $= s$ (in radians) $\times d$ (in km), which is:

$$R_{app} = 959.63 \times \frac{\pi}{180 \times 60 \times 60} \times 149\,597\,870.66 \ = 695\,992\,\text{km} \qquad (15)$$

which is the estimate given in the Ephemerids before the year 2000^{15}. One must say here that the "second of arc" is an astronomical unit a few odd, as nobody is able to get a mental representation to what this unit (and its divisions) specifically corresponds in the practical current way of life[16]. The reader will retain here that all recent publications made by solar American astronomers give solar dimensions in meters (or millions of meters –Mm– when they are speaking of layers below the solar surface), which is of an implacable logic as UAI has adopted the metric system (summit of the paradox, it is not rare that some French people says that this unit Mm is not meaningful, even though France is at the origin of the metric system).

As everyone knows, all diameters of stars are referenced by comparison to the solar diameter: for example $R\,(Antares) = 700\,R_\odot$ or $R\,(61\,Cygni\,A) = 0.8\,R_\odot)$. Thus, it does not seem harmless to know if the solar diameter is known to ±700 km. If that seems not too much significant for Antares, it is yet a little bit more for 61 Cygni A...

That is why a very nice experiment might be upgraded by amateurs, such people using today very efficient equipments. This concerns the measurement of the solar diameter as first made by Picard (1620–1682), an Abbot who was royal astronomer at the Paris Observatory (Fig. 19), and who is considered today as the pioneer of the solar astrometric measurements (that is to say the first man who realized accurate observations).

5.2 Meridian transit time of the Sun

The method consists in determining the needful time during which the Sun crosses a fixed line. To be rigorous, this process imposes an observation with a telescope (or a refractor) equipped with a reticule allowing to visualize the passages of the two diametrical limbs of the Sun; a precise clock provides the transit time.

At the end of the XVII th century, a small circle quarter was used, oriented in the meridian plan of the observation site, used jointly with a pendulum beating the second. It must be noticed the skill of the operators at that time working *with the eye and with the ear*. We may add that this method is never else than the ancestor of the modern technique called today *drift-scan method*.

---

[15] Allen, "Astrophysical Quantities", 3rd ed. gives 695 997 km. Note the inconsistency: adopting R = 695 508 km for the solar radius, then it is equivalent to 958.96'' and not to 959.63''.

[16] It is even worse when astronomers describe the atmospheric turbulence in this same unit, second of arc, while a unit like the cm – for which everyone has a tangible idea– perfectly characterizes the state of the atmosphere, through the Fried parameter, which is a physical parameter.

Jean PICARD (1620 - 1682)

Fig. 19. The Abbot Jean Picard, royal astronomer at the Paris Observatory (1620–1682). He is considered as the pioneer of astrometric solar measurements. Photograph taken from the web site *CNES-Picard*.

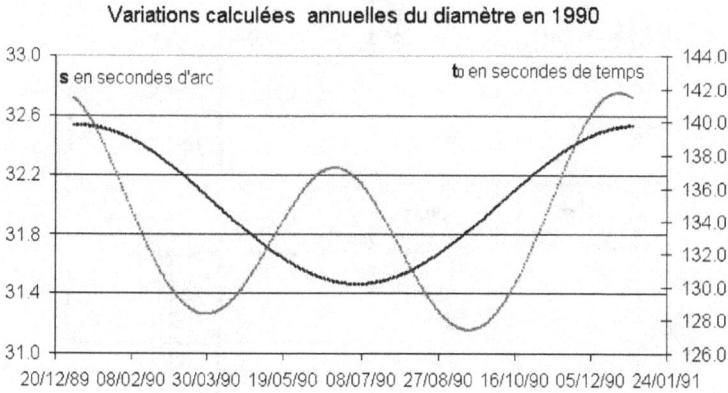

Fig. 20. Calculations for the year 1990 of the solar diameter (left scale) and meridian transit time (right scale). See Annexe 1.

The transit time τ_0 is given by (see Annexe 1):

$$\tau_0 = 127.947/(d \times \cos \delta) \quad \text{in} \quad \text{second} \quad \text{of} \quad \text{time}$$

where d represents the geometrical distance (in A.U.) of the Sun to the Earth and δ the declination of the Sun, each of them at the time of the observation[17].

---

[17]Note that these two quantities can be obtained rigorously at the instant of the observation, as modern ephemerides give d through Tchebichev polynomia, and δ can be accurately interpolated; such a precision is needed.

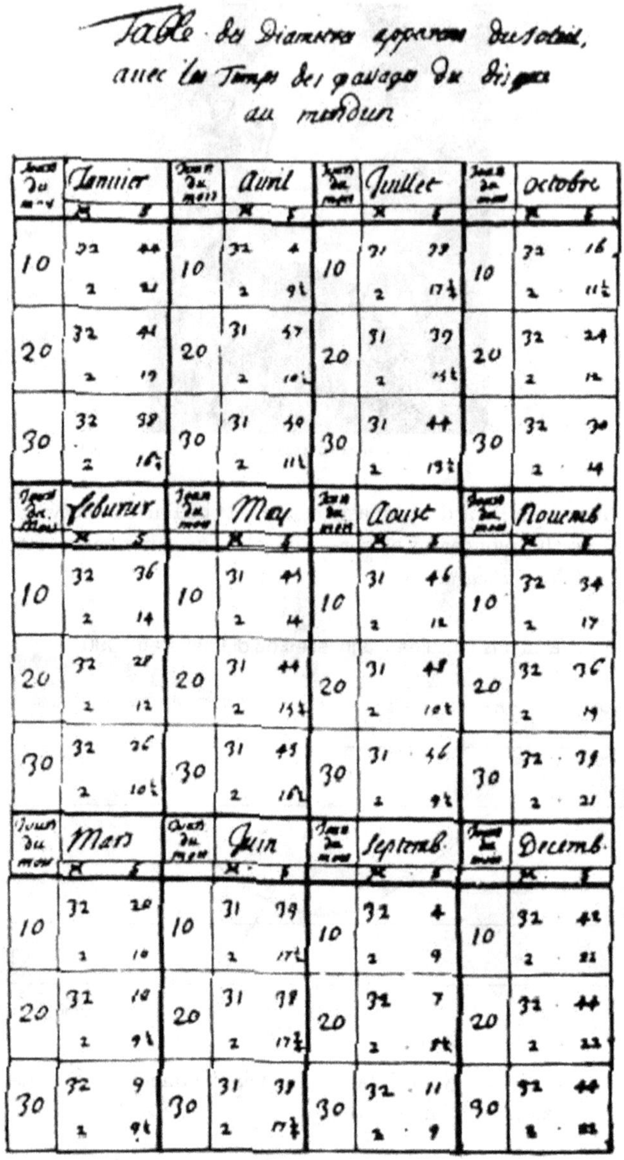

Table des Diamètres apparents du soleil, avec les Temps des passages du disque au méridien

Jour du mois	Janvier		Jour du mois	Avril		Jour du mois	Juillet		Jour du mois	Octobre	
10	32	44	10	32	4	10	31	33	10	32	16
	2	21		2	9½		2	17½		2	11½
20	32	41	20	31	57	20	31	39	20	32	24
	2	19		2	10½		2	14½		2	12
30	32	38	30	31	50	30	31	44	30	32	30
	2	16½		2	11½		2	13½		2	14

Jour du mois	Février		Jour du mois	May		Jour du mois	Aoust		Jour du mois	Novemb	
10	32	36	10	31	43	10	31	46	10	32	34
	2	14		2	14		2	12		2	17
20	32	21	20	31	44	20	31	41	20	32	36
	2	12		2	14½		2	10½		2	19
30	32	26	30	31	45	30	31	46	30	32	39
	2	10½		2	15		2	9½		2	21

Jour du mois	Mars		Jour du mois	Juin		Jour du mois	Septemb		Jour du mois	Decemb	
10	32	20	10	31	39	10	32	4	10	32	42
	2	10		2	17½		2	9		2	21
20	32	10	20	31	38	20	32	7	20	32	44
	2	9½		2	17½		2	9½		2	22
30	32	9	30	31	39	30	32	11	30	32	44
	2	9½		2	17½		2	9		2	22

Fig. 21. Manuscript archived in the Library of the Paris Observatory, relating observations of Villiard during the year 1676.

These terms are currently known with a great precision and are given for example by tables of the Paris *"Institut de Mécanique céleste et de calcul des éphémérides (IMCCE)"*. If M designates the mean anomaly (which stipulates that one looks to a mean distance at which orbits the Earth around the Sun), and e the orbital

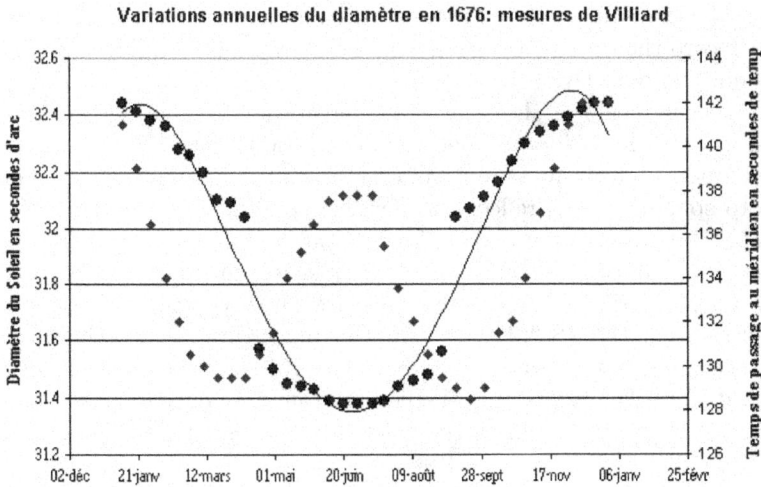

Fig. 22. Transcription under a modern form of the Villiard measurements, according to the Table shown in Figure 21. The precision is astonishing for such an old measurement by means of the transit time method. The dotted curve is a quadratic modern adjustment.

eccentricity, the diameter s of the Sun (neglecting terms of second order) is:

$$s = R_\odot \times (1 + e \times \cos(M))$$

M is calculated from the average motion n by $M = n \times (t - t_0)$; $n = 0.9856°/j$, t_0 is the date of the passage to the perihelion and t the date of the observation.

Figure 20 shows the duration variations of the meridian transit time and variations of the solar diameter in the course of the year 1990. Annexe 1 provides necessary calculations for celestial mechanics.

Thus, for the year 1990, the computations give the following transit times to the meridian: 128.350 s at the Spring equinox, 137.230 s at the Summer solstice, 127.417 s at the Autumn equinox and 127.775 s at the Winter solstice. The annual average is 133.549 s (or 2 mn 13 s 5).

It is remarkable that as soon as 1676, Etienne Villiard (who was first a scholar and then the assistant of Picard, according to the proper words of Villiard), has been able to measure the meridian transit times of the Sun at the Paris Observatory, according to this method. Figure 21 is extracted from the manuscript D-1, 17, p. 232 archived in the Observatory Library, while Figure 22 transcribes some data under a contemporary form. One will notice for example that the measurement made by Villiard on June, 20, 1676 is 137.75 s, an excellent precision considering the observational ways and means used at that time.

What famous predecessors did, young generations can do it nowadays, with a well better precision. It is not so much complicated: to put in evidence variations of 25 mas (angular millisecond of arcs) on the Sun diameter, it is accordingly necessary to record the time with a precision of the order of the millisecond (of time), which is quite easy today.

Such measurements of the solar diameter are far from being purely anecdotal. The theory shows indeed that shape variations must exist at the surface, due to the non uniform mass and velocity rate distributions. Albeit faint, distortions to sphericity thus reflect the physical conditions under the surface (the leptocline). Finally, temporal variations of the free boundary of the Sun, at a given altitude, involves astrophysical consequences, up to the level of the general relativity. Dedicated space missions are already scheduled for such purposes (for example the US "SDO" –Solar Dynamics Observatory, the successor of SOHO, successfully launched on February, 12th, 2010–, or the French-ESA mission "DYNAMICCS" –Dynamics and Magnetism from the Inner Core to the Corona of the Sun–), notwithstanding all the risks linked to such type of programs.

6 Conclusion

The shape of the Sun, as it appears to the naked eye, is an illusion; in other words, a perfect limb frontier does not exist. Indeed, the Sun is not a steel ball, but a gaseous sphere, and the passage from the "interior" to the "exterior" is gradually made. The boundary of this sphere has no proper characteristic, except if one decides, once for all, to take into account such or such physical parameter. To illustrate this purpose, it is commonly adopted today that the "edge" of the photosphere is delimited by the optical thickness $\tau = 1$. But one could say also that the solar disc is defined by all points of the surface where the temperature is minimal (see for instance this minimum on Fig. 4). One could say likewise that the disc is the place where the gravitational potential is equal to a constant $\phi = \phi_0$. This last definition has by far our preference, because the (gravity) level curves have a real physical meaning; they do not give a-priori constraints on the geometrical nature of the surface, as it is shown in Figure 17.

The entire spectrum of the Sun allows a complete description of our star, which gives us light and life. To conclude, it is a pity, that the sounds do not propagate in the empty space, because, moreover, we would hear the Sun playing all the day long a fantastic ode, if nevertheless, our ear was sensitive to all the emitted frequencies.

The author expresses its thanks to all readers that have helped to organize this document for a better understanding.

Annexe 1.   Measurement of the Sun's diameter: Meridian transit time method

Let be t_1 and t_2 the Western and Eastern contact time respectively, just at the moment where the Sun crosses the reticle of the eyepiece, due to the diurnal motion. The wire being oriented in the meridian plane, the apparent semi-diameter s is expressed as a function of the hour angle H of the Sun's center, between the two contact instants[18]:

$$H_1.\cos\delta = -s \text{ (at } t_1) \quad \text{and} \quad H_2.\cos\delta = +s \text{ (at } t_2). \tag{16}$$

In a first approach, one can neglect the δ declination variation between the two instants of contact, because $\cos(\delta)$ varies only of $1.0\,10^{-6}$ per minute of time (but this can be taken into account).

Let be TS the local sidereal time. The apparent diameter of the Sun can be written as:

$$2.s = \Delta H.\cos\delta \text{ with } \Delta H = \Delta(TS) - \Delta\alpha. \tag{17}$$

The passage duration $\tau_0 = t_2 - t_1$ is measured with a clock in a local scale of time. Let be k the coefficient allowing to convert sidereal time into the mean time (given in ephemerides, for example, for January 1, 1990, $k = 1.002\,737\,909$), then:

$$\Delta(TS) = k.\tau_0 \text{ in seconds of time.} \tag{18}$$

The variation $\Delta\alpha$ during the passage time is:

$$\Delta\alpha = k.\tau_0.\alpha' \tag{19}$$

where α' is the speed variation of α and is calculated through the daily variation $\Delta\alpha_{0,24}$ (in h ST –hour Sidereal Time) of the right ascension of the Sun between 0 and 24 h (UT) concerning the day where the observations take place:

$$\alpha' = \Delta\alpha_{0,24}/(24.k) \tag{20}$$

Hence

$$2.s = 15.k.\tau_0.(1 - \alpha').\cos\delta \tag{21}$$

s being expressed in seconds of arc (the factor 15 comes from the hours conversion of the hour angle H into seconds of time). The apparent radius R_{app} of the Sun is then deduced, and carry back to 1 AU, through $R_{app} = s.d$, that is:

$$R_{app} = (7.5).k.\tau_0.(1 - \alpha').d.\cos\delta \tag{22}$$

where d is in AU, τ_0 in seconds of time and R_{app} in seconds of arc.

   Order of magnitude. Nowadays, ephemerides can be found on web sites[19] which give, for each day, the geocentric distance, d (in AU), and can be interpolated to find d at the moment of the observations. It is the same for the other values k, α' and δ. The conversion of the sidereal time into the mean time k is of the order of $k = 1.002\,737\,909$ (seconds of time ST per seconds of time UT).

---

[18]See Danjon 1959, p. 70.
[19]See for instance the IMCCE site: `http://www.imcce.fr/imcce.php?lang=fr`.

A mean value of α' is such as α varies of $24\,h$ (ST) in 1 sidereal year, which is $1/(366.2422) = 0.002\,73$ (in fact α' varies from $0.002\,50$ to $0.003\,09$). Hence, by means of Equation (22):

$$\tau_0 \approx 127.947/(d.\cos\delta) \quad \text{in seconds of time} \tag{23}$$

Computation in the case of a Keplerian orbit. Classical notations are used and numerical values are given, as an example, for the year 1990:

$l_\odot$	solar ecliptic longitude	$l_\odot = v + \omega + 180°$
v	true anomaly of the Earth	$v = M + 2e.\sin(M) + \frac{5}{4}.e^2.\sin(2M)$
ω	perihelion longitude	$\omega = 102.8°$
M	mean anomaly	$M = n.(t - t_0)$
n	mean motion	$n = 0.98561°/j$
t	date	t varies from 0 to 365 days
t_0	perihelion passage date	$t_0 \approx 3d.$
ε	obliquity of the ecliptic	$\varepsilon = 23.44°$

Using the relations describing a Keplerian orbit:

$$d = (1 - e^2)/(1 + e.\cos v) \quad \text{and} \quad \sin\delta = \sin\varepsilon.\sin l_\odot \tag{24}$$

one may also obtain an approximate value (from Eq. (22)) of R_{app} as a function of the ecliptic longitude of the Sun, allowing to check quickly the order of magnitude (because $k.(1 - \alpha') \approx 1)$):

$$R_{app} \approx 7,5.\frac{1 - e.\cos(l_\odot - \omega)}{\sqrt{1 - \sin^2\varepsilon.\sin^2 l_\odot}}.\tau_0. \tag{25}$$

This formula was used at the beginning of the XIX th century to tabulate the solar diameter values (see J.B. Delambre, *Astronomie théorique et pratique*, 1814, tome 2, ch. 24), as a function of the passage time τ_0.

Nowadays, due the easiness of modern computers calculation, it is straightforward to deduce R_{app} rigorously, especially using developments of v.

Finally, if one set an arbitrary absolute value R_f, let say for instance, 959.63″, then, to first order:

$$R = R_f(1 + e.\cos M + \mathcal{O}(M)) \tag{26}$$

which allows to deduce the variation of R along the course of the year (Fig. 20).

References

Cox, A.N., 2000, Allen Astrophysical quantities, Fourth edition (Springer ed.)

van Belle, G.T., Ciardi, D.R., Thompson, R.R., Akeson, R.L., & Lada, E.A., 2001, ApJ, 559, 1155

Dabrowski, B., & Benz, A., 2008, "The Good Guys and the Rascals", http://sprg.ssl.berkeley.edu/~tohban/nuggets/?page=article&article_id=86, October, 13, 2008

Beckers, J.M., 1969, "A Table of Zeeman Multiplets, Sacramento Peak Observatory". Technical Report, AFCRL-69-0115

Danjon, A., 1959, Astronomie Générale, Sennac ed., 446

Di Mauro, & Pia, M., 2003, "Helioseismology: a Fantastic Tool to Probe the Interior of the Sun", In "The Sun's Surface and Subsurface", Lecture Notes in Physics, Vol. 599, (Springer ed.), 31

de Souza, Domiciano, A., et al., 2003, A&A, 407, L47

Floyd, L., 2003, "Solar Ultraviolet Irradiance: Origins, Measurements and Models". In: "The Sun's Surface and Subsurface", Lecture Notes in Physics, Vol. 599 (Springer ed), 109

Foukal, P., & Vernazza, 1979, AJ, 267, 863

Fröhlich, C., & Finsterle, W., 2001, Astron. Soc. Pac. Conf. Ser., 203, 105

Godier, S., & Rozelot, J.P., 2000, A&A, 355, 365

Gelly, B., 2003, "Detection of Eigenmodes in Helioseismology", in: "The Sun's Surface and Subsurface", Lecture Notes in Physics, Vol. 599 (Springer ed.) 68

Habereiter, M., Schmutz, W., & Kosovichev, A.G., 2008, AJ, 675, L53, ed. J.P. Rozelot (First approach: 2002, Atelier de travail d'Annecy)

Howe, R., Christensen-Dalsgaard, J., & Hill, F., 2000, Science, 287, 2456 and Howe, R., 2009, "Dynamic Variations at the Base of the Solar Convection Zone", Living Rev. Solar Phys., 6

Hudson, H., & Rozelot J.P., 2010, RHESSI science nugget: http://sprg.ssl.berkeley.edu/~tohban/wiki/ index.php/History_of_Solar_Oblateness

Kitchin, C., 2001, "Solar Observing Techniques" (Springer ed.), 210

Lang, K.R., 1995, "Le Soleil et ses relations avec la Terre" (Springer ed.), 276

Lefebvre, S., & Rozelot, J.P., 2003, Astron. Soc. Pac. Conf. Ser., 286, 84

Lefebvre, S., Kosovichev, A.G., & Rozelot, J.P., 2007, ApJ, 658, L135

PICARD website: http://smsc.cnes.fr/PICARD/Fr/

Pireaux, S., & Rozelot, J.P., 2003, Astrophys. Space Sci., 284, 1159

Rondi, A., http://astrosurf.com/rondi/obs/shg/index.htm and 2007, in: "L'Astronomie"

Rozelot, J.P., Lefebvre, S., & Desnoux, V., 2003, Solar Phys., 217, 39

Rozelot, J.P., 2006,"Advances in Understanding Elements of the Sun-Earth Links", in: "Solar and Heliospheric Origins of Space Weather Phenomena", Lecture Notes in Physics, 699 (Springer ed.), 166

Rozelot, J.P., 2009, "What Is Coming: Issues Raised from Observation of the Shape of the Sun", in: "The Rotation of Sun and Stars", Lecture Notes in Physics, Vol. 765 (Berlin: Springer), 15

Séguin, M., & Villeneuve, B., 1995, Astronomie, Astrophysique (Masson, ed.), 547

Schroeder, D., 1987, Astronomical Optics (Academic Press Inc. ed.), 352

Sofia, S., & Li, L., 2001, Geophys. Res., 106, 12969

Stix, M., 2003, "The Sun", 2nd edition (Springer ed.), 390

Wilson, L.T., & Hüttemesister, 2000, "Tools of Radio Astronomy", 2 volumes; First, 440 p. and second, Problems and Solutions, 162 (Springer, ed.)

Astronomical Spectrography for Amateurs
J.-P. Rozelot and C. Neiner (eds)
EAS Publications Series, **47** *(2011) 139–163*

SPECTROSCOPY OF BE STARS

C. Neiner[1]

Abstract. This chapter describes non supergiant B-type stars that show emission lines, called Be stars. The emission is caused by the presence of a circumstellar decretion disk. Many physical phenomena are thought to be involved in these stars, such as rapid rotation, pulsations and magnetic fields, and give rise to variations. Spectroscopy is used as a diagnostic tool to study Be stars, by professional astronomers as well as by amateurs.

1 Introduction

The definitions of B and Be stars are given in Section 2, together with the definition of the "Be phenomenon". The typical spectrum of a Be star is described in Section 3 for each domain of wavelengths: optical, infrared (IR) and ultraviolet (UV). The polarisation of the spectral light is also considered. Section 4 summarizes the special cases of Be stars: Be-shell stars, B[e] stars and Be stars in binaries.

An usual property of Be stars is their rapid rotation. This property has strong effects on the star and its spectrum, which are described in Section 5. Be stars are also strongly variable stars at all timescales, as shown in Section 6. These variations are thought to be related to pulsations (Sect. 7) and maybe magnetic fields (Sect. 8).

Be stars are a fascinating class of stars for professional astronomers as well as for amateurs. Several scientific programs for the study of Be stars are already undertaken by amateurs and are summarized in Section 9. Advices for amateurs who would like to observe Be stars are given.

Finally, conclusive remarks are drawn in Section 10.

---

[1] LESIA, UMR 8109 du CNRS, Observatoire de Paris-Meudon, 5 place Jules Janssen, 92195 Meudon Cedex, France

© EAS, EDP Sciences 2011
DOI: 10.1051/eas/1147005

2   Definitions

2.1   B-type stars

B-type stars are hot and blue stars, with a high proportion of radiation in the ultraviolet domain. Their temperature varies from about $10\,000\,K$ at the subtype B9 to nearly $30\,000\,K$ at B0. Their mass ranges from about 3 to $20\,M_\odot$ and their luminosity from 100 to $50\,000\,L_\odot$.

Moreover, B-type stars are often found together with O stars in OB associations since, being massive, they are short-lived and therefore do not survive long enough to move far from the place where they were formed. Spica and Regulus are two well-known examples of B-type stars.

The spectrum of B stars is characterized by the presence of hydrogen and HeI lines in the optical. The strength of hydrogen increases with the subtypes (from B0 to B9), while the maximum intensity of the HeI lines is reached around the subtype B2 (see Fig. 1). HeII lines only appear in the B0 stars. Other spectral features may include CaII, CII, CIII, NII, NIII, OII, SiII, SiIII, SiIV, and MgII.

Variability is found in several classes of B stars, for example in β Cephei, Slowly Pulsating B (SPB) and Be stars.

2.2   Be stars

Be stars are non-supergiant B stars that at least once have displayed Balmer line emission (Collins 1987, and see Fig. 1). In the stellar classification, the "e" in "Be" stars thus stands for "emission". This property applies to about 20% of all B-type stars in our galaxy, but is especially observed at the B1-B2 subtypes. Some late O and early A stars also show such emission and are considered as an extension of the Be stars. The early (*i.e.* more massive and hotter) Be stars exhibit strong variable winds evidenced by the rapidly variable UV resonance lines of highly ionised species (*e.g.* CIV or SiIV), as well as by spectral and light variations in the optical on timescales from hours to decades.

The visual inspection through a spectroscope of the first discovered Be star, γ Cas, was described in 1867 by Secchi. Other bright examples of Be stars are ζ Tau or δ Sco.

2.3   The Be phenomenon

The phases of emission in the optical and infrared lines of hydrogen and several ions, called the Be phenomenon, most likely reflect changes in the structure of a circumstellar disk created by episodic ejections of mass. The typical mass-loss rate of Be stars is about $10^{-8}\,M_\odot\,\mathrm{yr}^{-1}$. The origin of this phenomenon is, however, still unknown.

The key questions in understanding this phenomenon are:

- how to generate the angular momentum needed to eject material so that it can attain a stable orbit around the star?

- how to eject these quantities of mass in the observed episodic fashion?

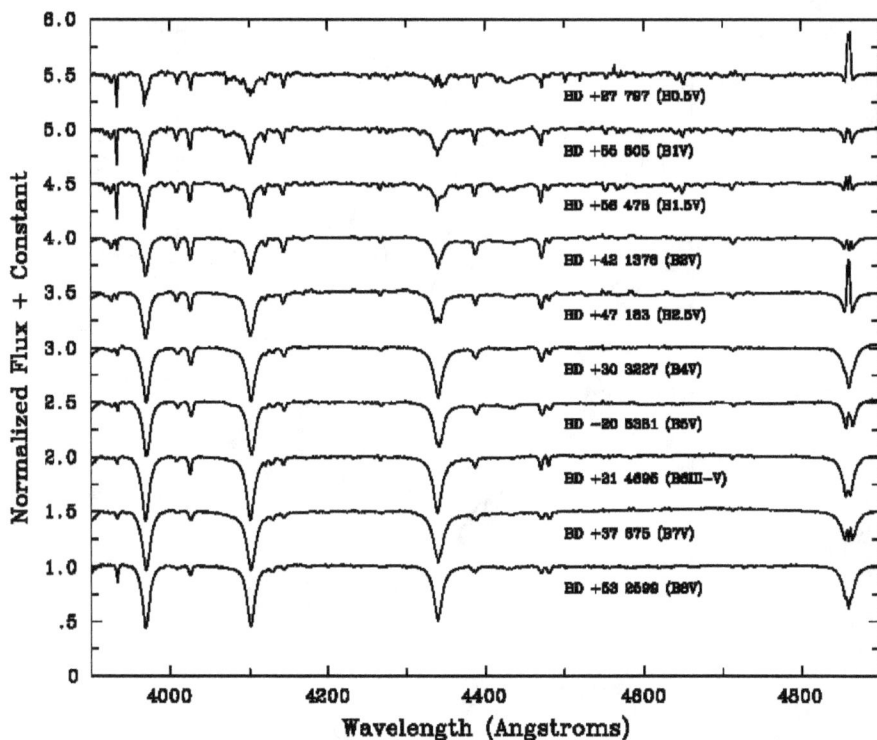

Fig. 1. Examples of main-sequence Be star spectra. H lines are seen at 3970 (Hε), 4101 (Hδ), 4341 (Hγ) and 4861 Å (Hβ). Note the progressive decline in the intensity of the HeI lines (4009, 4026, 4471 and 4713 Å) from a maximum at B1-B2 subtypes and the increase of the MgII line at 4481 Å with the spectral type. Also note the emission in Hβ, which makes these B stars Be stars. Taken from Steele *et al.* (1999).

A few explanations have been proposed: beating of non-radial pulsations (NRP) modes, presence of a magnetic field, hot spots, photospheric shocks and/or mass transfer in interacting binaries (see *e.g.* Baade 2000; Balona 2000). Up to now, however, observational facts and theories did not converge towards a unique conclusive explanation.

3   The spectrum of a Be star

3.1   The optical spectrum

The Balmer lines (Hα, Hβ, etc.) are the ones most commonly seen in emission in Be stars. The level of emission decreases along the Balmer series, the Hα line having the strongest emission. Moreover, lines from other ions can also be in emission, in particular the HeI lines at *e.g.* 4921, 5876 and 6678 Å, SiII at 6347 Å,

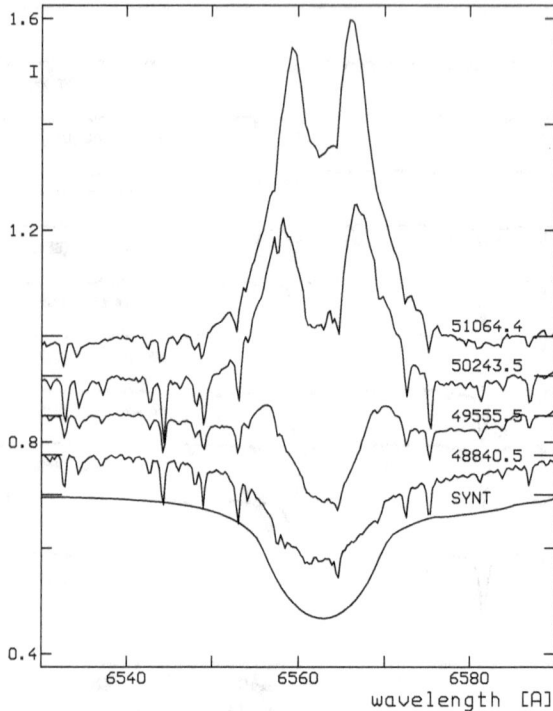

Fig. 2. Examples of Hα profiles of the Be star 60 Cyg in the non-emission, intermediate and emission phase (Koubský *et al.* 2000). The bottom spectrum is a synthetic profile of this star. The upper four spectra were obtained at different times over a 6 year interval, between HJD 2448 840.5 and 2451 064.4.

MgII at 4481 Å, CII at 6578 and 6583 Å and the numerous FeII lines. The emission level is highly variable from one Be star to the other, but also in time for a given Be star: it can even completely disappear and reappear years later.

Figure 2 shows examples of Hα spectra of the Be star 60 Cyg taken at different times, compared to a synthetic spectrum for a star of similar spectral type. This star goes from a phase with the Hα line in absorption to a phase with strong double-peaked emission within 6 years.

The presence of a circumstellar disk, seen under different angles depending on the star, gives rise to different kinds of emission line profiles. When the star is seen pole-on, *i.e.* from the top, the equatorial disk is seen face-on and the line has a single emission peak. When the star and the disk are seen edge-on, *i.e.* from the equator, the line has a double-peak emission with a central absorption. The two emission peaks are then called the Violet (V) and Red (R) peaks. Any line of sight with an angle between 0 and 90° will give a line profile with a mixture of emission and absorption. Examples are shown on Figure 3. Moreover, if the disk is structured as several successive rings, the emission profile is the sum of

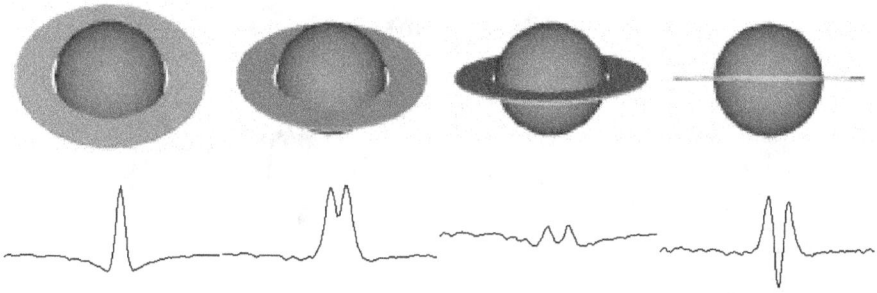

Fig. 3. Examples of line profiles depending on the inclination angle between the observer's line of sight and the axis of the disk. The last example is a typical Be-shell case (see Sect. 4.1). Rotational model taken from Slettebak (1979).

the emission of each ring. One can then observe *e.g.* four emission peaks instead of two.

The presence of a circumstellar disk not only affects the lines but also the continuum. While B stars show one Balmer discontinuity at 3700 Å, Be stars show two Balmer discontinuities: the one at 3700 Å and another one at 3647 Å. This second discontinuity can be seen in emission when the star is in a phase either of strong emission or of absorption in the case of Be-shell stars (see Sect. 4.1). This additional discontinuity is due to the presence of the circumstellar enveloppe. Figure 4 shows examples of these discontinuities.

3.2 In the UV and beyond

In the UV, the spectral energy distribution of Be stars is similar to the one of B stars. However, Be stars have large blue-winged resonance lines, such as CIV (1550 Å), SiIV (1400 Å), NV (1240 Å) or AlIII (1860 Å), indicative of an enhanced stellar wind (Grady *et al.* 1989).

Most of the single Be stars show a X-ray luminosity similar to B stars but slightly higher (Cohen *et al.* 1997). This is coherent with the presence of a disk around Be stars. Moreover, Be stars in binaries have a strong X-ray luminosity, showing strong brightenings on timescales of weeks to years (see Sect. 4.3).

3.3 In the IR and beyond

In the IR, Be stars are characterised by an excess of light compared to B stars. This is due to the presence of the circumstellar disk.

The Paschen continuum of Be stars can be brighter than the one of B stars. The spectra are dominated by free-free and free-bound emission from the disk in the near- to mid-IR region. The spectral energy distribution in the far-IR and radio domains indicates structural changes far away from the central star (Waters *et al.* 1991).

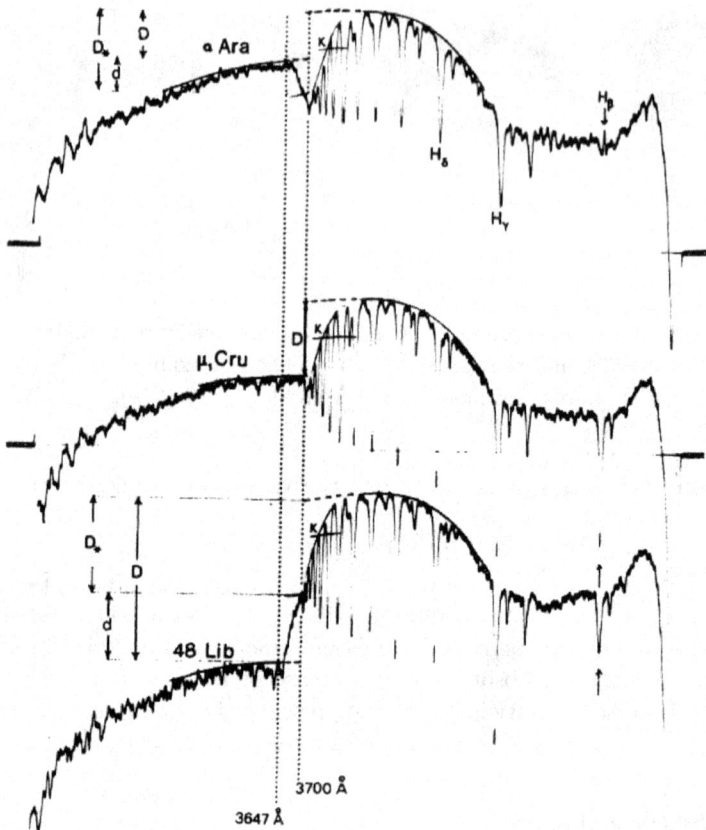

Fig. 4. Observations of the Balmer discontinuities. The B star μ^1 Cru shows only one discontinuity at 3700 Å. The Be stars α Ara and 48 Lib show a second discontinuity at 3647 Å due to the presence of the circumstellar enveloppe (Divan 1979).

Paschen lines are found in emission in Be stars. They are strong in early Be stars (B0-B3) and decrease in intensity towards late Be stars (Andrillat *et al.* 1990). Be stars with Paschen lines in emission also show a greater excess of flux in the near-IR domain than stars with Paschen lines in absorption (Briot 1977).

Some B[e] stars (see Sect. 4.2) also show the presence of dust, probably remnants from earlier phases of the evolution of the star.

3.4 Polarisation of the light

Almost all Be stars emit polarised light in their continuum. This is due to the presence of the disk. The amount of polarisation (up to 2%) is proportional to the emission level of the star, but time lags have been sometimes observed (Poeckert *et al.* 1979). Polarisation strength may also vary with other

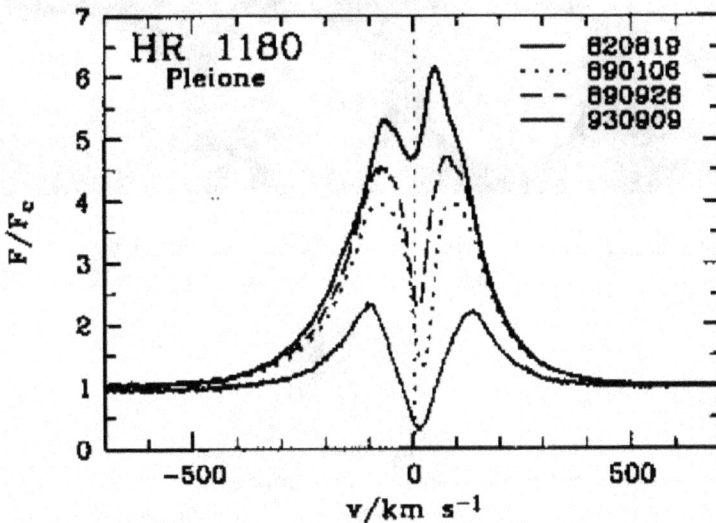

Fig. 5. Hα observations of Pleione (Hanuschik *et al.* 1996). Note the transition from Be-shell spectrum (until 1989) to Be spectrum (1993).

quantities, such as the V/R ratio (see Sect. 6.2), while the polarisation angle stays constant.

Finally, polarisation in the lines can also be detected if the star hosts a magnetic field. This may be the case of the star ω Ori (Neiner *et al.* 2003, and see Sect. 8.3).

4   Special Categories of Be Stars

4.1   Be-shell stars

Be-shell stars are Be stars with deep and narrow absorption lines in their spectra, for example sharp absorption lines of FeII, TiII and CrII (see Fig. 3). Most of the time they are stars with a large $v \sin i$ which broadens the lines (see Sect. 5.2), thus the narrow absorptions cannot come from the star but from a dense equatorial circumstellar region where the optical depth is large. Note that this shell property can be temporary. The stars Pleione and V 1924 Aql are examples of Be stars that sometimes show shell lines (see Fig. 5), *i.e.* changing from Be stars to Be-shell stars.

4.2   B[e] stars

B[e] stars are Be stars that exhibit forbidden emission lines in their spectra (*e.g.* [OI] at 6300 Å or [NII] at 6583 Å). These lines are often attributed to supergiants, which have lower density. B[e] stars also show IR excess, larger than those of Be stars, due to circumstellar dust emission.

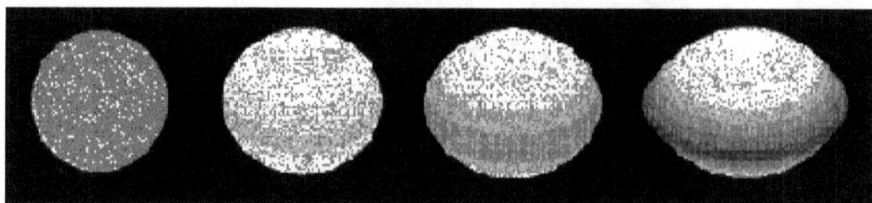

Fig. 6. Rotational flattening of the star, due to the centrifugal force. The rotation velocity increases from left to right. The more the star rotates rapidly, the less it is spherical.

4.3 Be stars in binaries

About one third of the Be stars are in binaries. In case of a wide binary system, no interaction between the Be star and the companion is possible. However, in case of a close system, the companion can have a strong effect on the Be star. In particular, the tidal forces may help the Be star to eject material. The companion might also have helped the Be star to spin up. The case of Be stars in binaries should therefore be considered as a special case and the physical processes in these binaries cannot be compared to the ones taking place in single Be stars.

5 Rapid rotation

Be stars usually rotate fast. Slow examples are known *e.g.* β Cep with v = $28 \, \text{km s}^{-1}$, but are rather rare and often considered as special cases. Therefore, rapidly rotating Be stars are also called "classical" Be stars.

Rapid rotation has a strong effect on the star itself (oblateness), as well as on its spectrum (Doppler broadening). These effects are described below.

5.1 Oblateness

The faster a star rotates, the more it is flattened by the centrifugal force and the less spherical it is (see Fig. 6). If the star rotates fast enough, material will even flow away from the star at the stellar equator. Struve (1931) thus proposed that the origin of the circumstellar disk of Be stars is the high rotation rate of these stars.

It is indeed generally accepted that the envelope of Be stars is flattened by their high rotational velocities. Interferometric observations (*e.g.* Quirrenbach *et al.* 1997; Stee 2000) provided direct evidence for the presence of such an asymmetry. Recently, Domiciano de Souza *et al.* (2003) presented the first observation of a Be star, α Eri, with the VLTI (ESO). The results show that this star is not at all a sphere but is oblate (see Fig. 7).

However, the rotation rates of Be stars are always lower than the critical velocities at which the centrifugal force balances gravitation at the equator, *i.e.* at which material can escape from the star. On average, the Be stars rotate at about

North

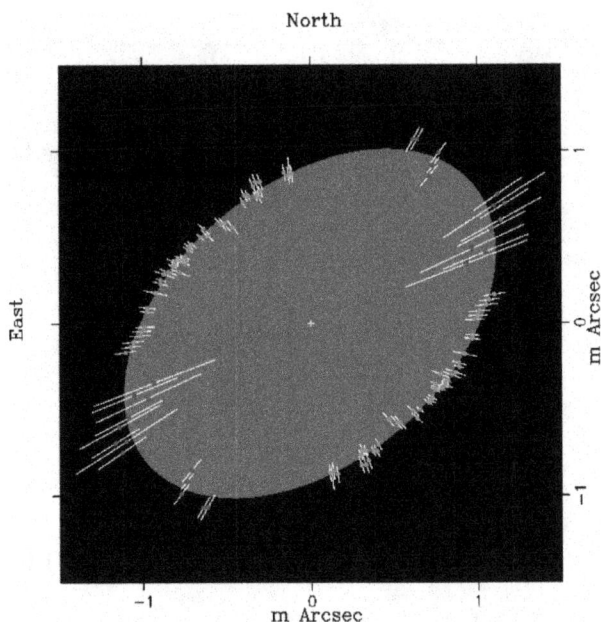

Fig. 7. Interferometric measurements of α Eri obtained with the VLTI by Domiciano de Souza *et al.* (2003), showing that this star is oblate.

80–90% of the critical velocity (Chauville *et al.* 2001; Frémat *et al.* 2005). Note, nevertheless, that the rotation rates of Be stars have been recently questioned by Townsend *et al.* (2004). These authors think that Be stars could rotate at 95% of the critical velocity. Still, the centrifugal force by itself is inadequate to explain the formation of a disk around these stars and another mechanism has to be invoked, such as stellar pulsations.

5.2 Doppler broadening

When a star rotates, the different parts of the star have a different velocity component towards the observer. Therefore, a line is not only formed at its rest wavelength, but on a range of wavelengths around the rest wavelength. This is called "Doppler broadening" (also see Chapter 1). As a result, the spectral line is broader, and each part of the line corresponds to a certain place on the surface of the star (Fig. 8).

The inclination angle i at which the star is seen also influences the width of the line profile. Whether the star is seen pole-on or equator-on will make the lines narrow or broad respectively. Therefore $v \sin i$ is commonly used for the velocity of Be star instead of v itself. $v \sin i$ can be easily determine, *e.g.* with a Fourier analysis of the line profile: the first zero of the Fourier transform is then $v \sin i$. The value of the velocity v itself, however, can only be determined with models of stellar atmospheres.

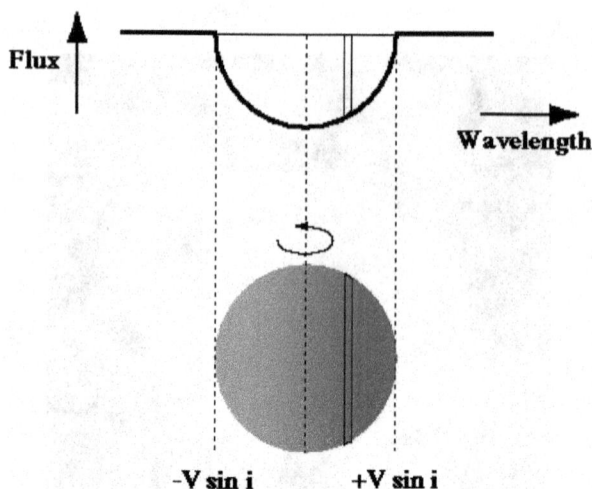

Fig. 8. Rotational broadening of the line profiles, due to the Doppler effect.

6   Variations

Be stars undergo variations at all timescales, related to different physical phenomena. First of all, rapid periodic variations are related to pulsations and rotation. Moreover, long-term variations are associated to the wind and disk. Finally, ejections of material from the star into the disk produce sudden episodic variations.

6.1   Rapid variations

Periodic variations of the order of one day occur in Be stars, either in the photosphere or in the immediate circumstellar environment. They are induced by pulsations (see Sect. 7) and rotational modulation. They are observed in photometry and in the line profiles (Fig. 9) of most early Be stars. They are also seen in the radial velocity, equivalent widths and central depths variations. Figure 10 shows the rapid variations of the radial velocity of the Be star 66 Oph with the pulsation frequency.

However, the search for rapid variations in late (B5 and later) Be stars is less easy. This is due to the detection limit with present ground instrumentations but can be investigated with the space missions such as CoRoT.

6.2   Long-term variations

Variations on a longer timescale are also observed.

Cyclic variations of the order of months or years are associated to the wind. The wind of Be stars is stronger than the one of B stars and variable. This can be observed in particular in the UV resonance lines.

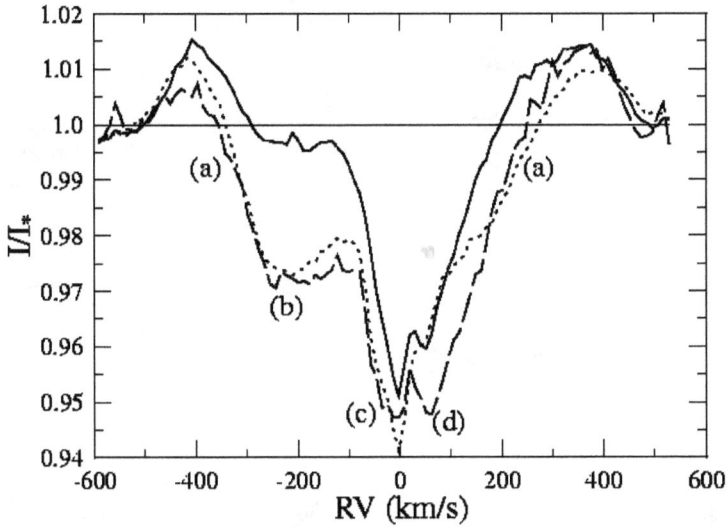

Fig. 9. Rapid variations observed in EW Lac over 5 days. Residuals are shown, *i.e.* each spectrum has been divided by the average spectrum, to let the differences in the line profile appear. The dotted line spectrum was taken on day 1, dashed line spectrum on day 2, full line spectrum on day 6, in August–September 1993 (Floquet *et al.* 2000).

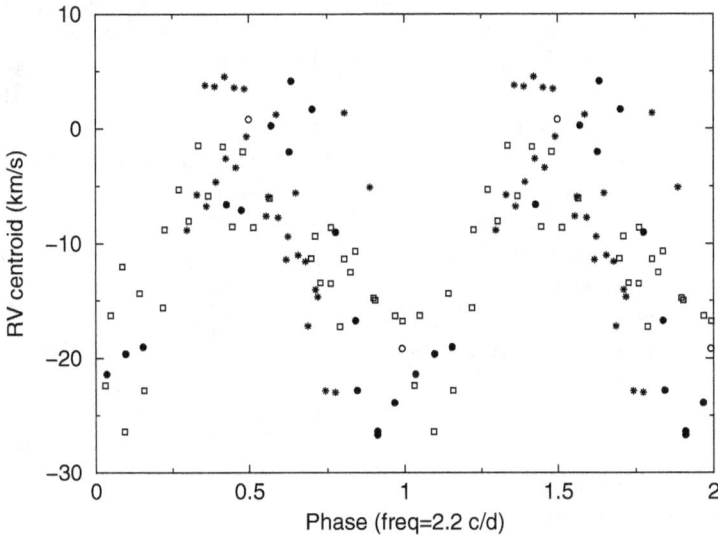

Fig. 10. Radial velocity variations of the centroid of the HeI line at 6678 Å, folded with the pulsation frequency $f = 2.2$ c d^{-1}, for observations of 66 Oph obtained in 1997 (filled circles), 1998 (open squares) and 2001 (stars and open circles for LNA and TBL observations respectively). Taken from Floquet *et al.* (2002).

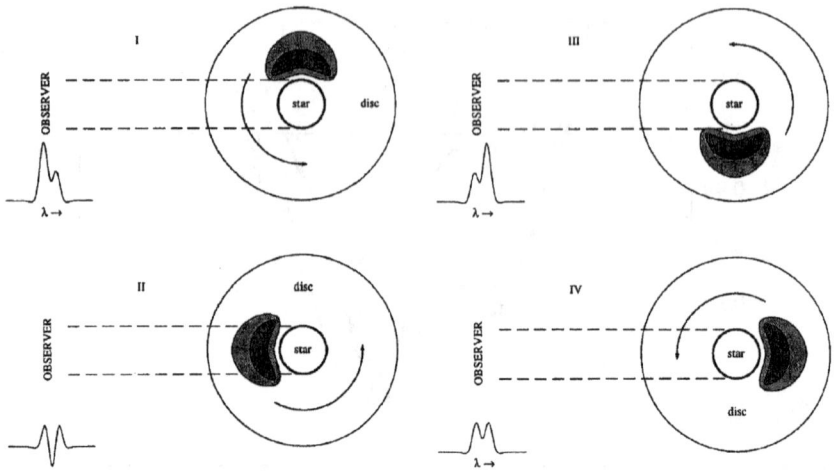

Fig. 11. One-armed oscillations in the disk of Be stars (Telting 1997). A denser zone in the disk passes in the observer's line of sight, changing the intensity of the V and R peaks in the emission profile.

On the other hand, the intensity of the violet (V) and red (R) peaks of double-peak emission lines varies with a timescale of a few years to decades. The distribution of timescales of the V/R variation has its maximum around 7 years and no spectral-type dependency is found (*e.g.* Hirata & Hubert-Delplace 1981). This behaviour is consistent with one-armed oscillation in the disk. This global disk oscillation is due to a denser zone in the disk, which slowly precess around the star, thus passing in the line of sight of the observer in the prograde direction, *i.e.* in the direction of rotation (Telting 1997). This has been confirmed by interferometric measurements, *e.g.* by the movement of the photocenter of the Hα emission of ζ Tau (Vakili *et al.* 1998). When the dense part is approaching the observer, the V peak is stronger, whereas when the dense part is going away from the observer, the R peak is stronger (Fig. 11).

Moreover, cyclical radial velocity variations are associated to the V/R variations and are typical for early Be-shell stars (*e.g.* ζ Tau, 48 Lib, EW Lac).

Furthermore, the intensity of emission lines slowly varies in time (see Fig. 2). The timescale of variation of the emission line intensity relative to the continuum level (I_{max}/I_c) is generally shorter in early Be stars than in late Be stars (*e.g.* Hirata & Hubert-Delplace 1981). Figure 12 shows the long-term evolution of the emission intensity in Hα for the Be star ω Ori.

6.3 Episodic ejections of material

In addition to the short-term periodic and long-term cyclical variations, ejections of material episodically occur, with a timescale of a few months to decades. The material that is ejected in this way goes into a stable orbit around the star, thus

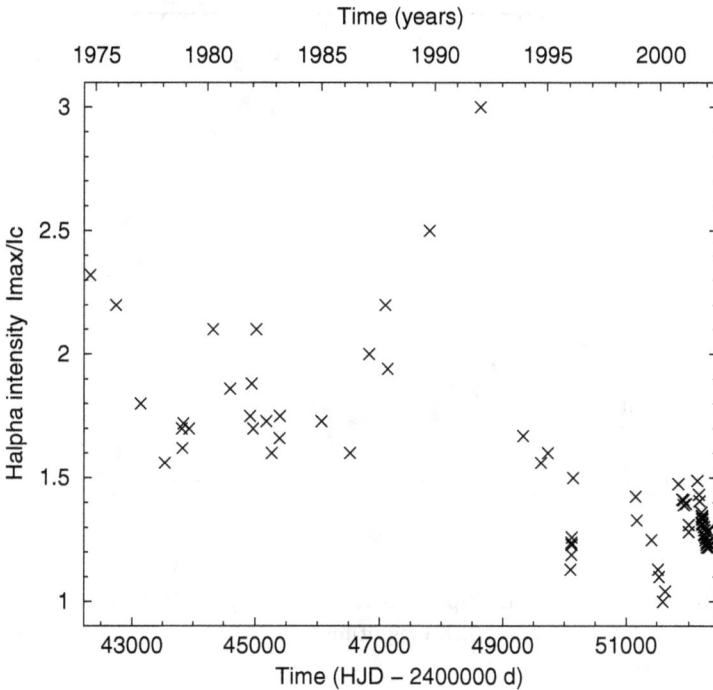

Fig. 12. Long-term evolution of the normalized Hα intensity of ω Ori (Neiner *et al.* 2002). Data are taken from Andrillat & Fehrenbach (1982), Balona *et al.* (2001), Banerjee *et al.* (2000), Bopp & Dempsey (1989), Buil (2001), Dachs *et al.* (1981), Doazan *et al.* (1991), Hanuschik *et al.* (1996), Oudmaijer & Drew (1999), Sonneborn *et al.* (1988), Srinivasan (1996) and Neiner *et al.* (2002).

creating the disk. The disk can sometimes completely disappear and reappear years later.

When the disk is building up, a brightening is observed in the V magnitude (Fig. 13), which is called an "outburst". At the same time, the emission builds up in the lines. For Be stars with a dense disk, the additionnal ejected material can make the disk so dense that no emission can exist anymore. In that case, instead of an increase of the emission in the lines during the outburst, one can see the emission completely disappear.

7   Pulsations

During the last ten years, stellar pulsations have been put forward as a possible mechanism to explain the Be phenomenon. The theory and observations of pulsations are presented in this section.

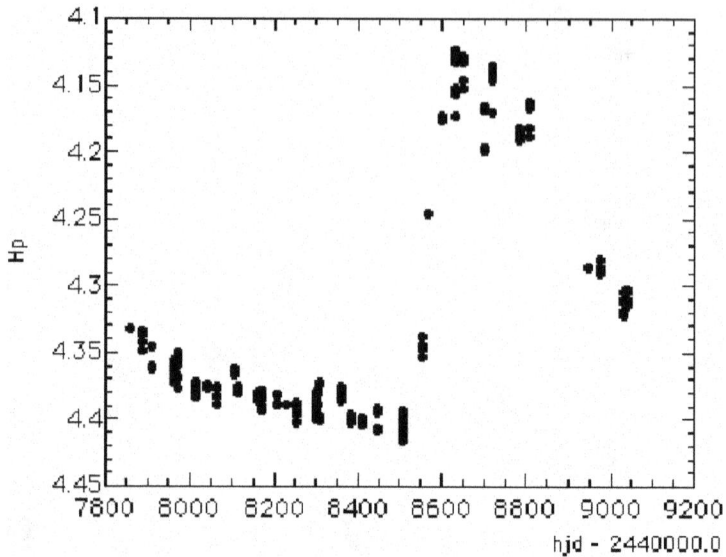

Fig. 13. Outburst observed in Hipparcos data of v Cyg by Hubert & Floquet (1998). A strong increase of brightness (0.25 magnitude) is observed over 100 days, followed by a slow gradual decrease (>400 d).

7.1 Theory

Oscillation patterns observed at the surface of stars are caused by sound waves that propagate into the interior of these stars. A sound wave travels from the surface almost straight towards the centre of the star. Its path then slowly bends around, because of the increasing sound speed, so that it misses the centre of the star. How exactly it propagates depends on the details of the sound speed inside the star. The point of closest approach is known as the turning point of the mode. After the turning point the wave moves out again until it reaches the surface. At the surface it is reflected as if by a mirror and it goes back deeper in the star again.

Since each of the modes follows a slightly different path through the interior of the star, it senses the sound speed in slightly different parts of the interior. Using a mathematical inversion it is possible to use very slight differences in frequencies of different pulsation modes to deduce the internal structure of the star. This technique is called asteroseismology.

Two types of stellar oscillations are distinguished: radial and non-radial. The radial pulsation is the simplest type in which a star oscillates around its equilibrium state by expanding and contracting in a periodic way, while keeping its spherical shape. In the case of general non radial pulsations (NRPs) the total volume is conserved, but not the sphericity, as neighboring segments of the star oscillate in different phases. Theoretically, radial pulsations can be treated as a special case of NRPs.

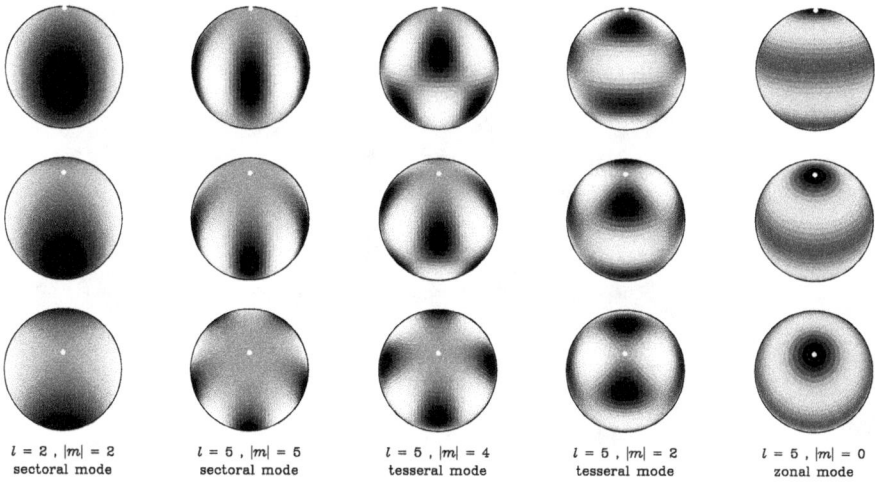

| $l = 2$, $|m| = 2$ | $l = 5$, $|m| = 5$ | $l = 5$, $|m| = 4$ | $l = 5$, $|m| = 2$ | $l = 5$, $|m| = 0$ |
| sectoral mode | sectoral mode | tesseral mode | tesseral mode | zonal mode |

Fig. 14. Examples of non-radial pulsations modes (Schrijvers 1999). Black areas move outward, whereas the white areas move inward.

NRP modes are characterised by three main parameters: the pulsation degree ℓ, the azimuthal order m, and the radial order n. The stellar surface is divided into parts oscillating in antiphase defined by ℓ border lines, of which $|m|$ are spaced equally in azimuth for each constant longitude. Thus $\ell - m$ is the number of border lines of equal latitude and $-\ell \leq m \leq \ell$. n is the number of nodes from the center to the surface of the star. Examples of pulsation modes are shown in Figure 14. For further reading and mathematical formulations, the book by Unno *et al.* (1989) is recommended.

It should be noted that pulsation periodicities of rotating stars are difficult to predict by theory, because many physical effects cannot be taken into account yet in the theory of stellar structure, especially in case of rapid rotation. Moreover, in a real star, the amplitude of pulsations is limited by non-linear processes, which cannot be calculated yet either.

7.2 Pulsations in Be stars

The velocity of NRPs on the stellar surface is very small. Therefore, NRPs cannot, by themselves, eject matter from the star. However, NRPs could feed energy into the equatorial surface layers of the star to accelerate them to the critical rotational velocity. It is then the centrifugal force, *i.e.* the rotation, which would be the final cause of the mass ejection, but pulsations would be the cause for critical rotation velocity.

When several pulsation modes are present, constructive interferences called a "beating" can occur. In that case, the amount of energy fed to the surface layers at the moment of beating is higher than for a single pulsation mode and thus the chance of reaching the critical velocity and an ejection of material is higher.

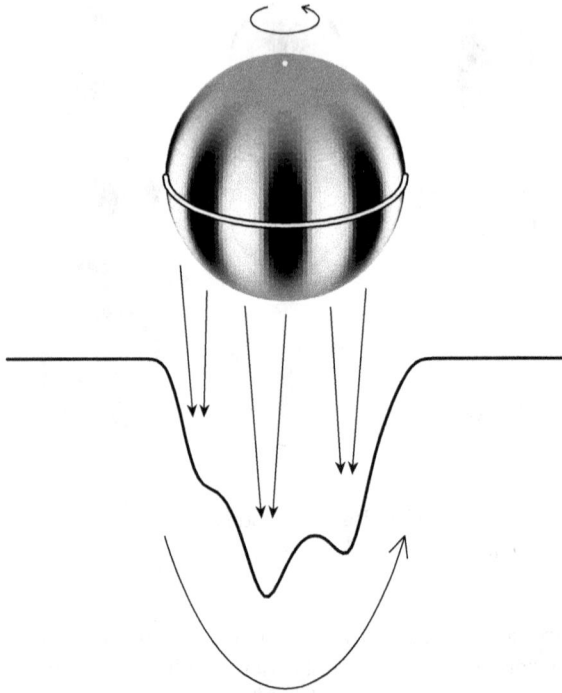

Fig. 15. Schematic line-profile deformations, due to non-radial pulsations, moving along the line profile as the star rotates (Schrijvers 1999; after Vogt & Penrod 1983).

7.3 Observations

NRPs are expected to occur in basically all stars, but it is difficult to find them, because NRPs can only be detected as very small perturbations moving through the line profiles, affecting often less than 0.1% of the line intensity (see Fig. 15). This requires 2m-class telescopes equipped with high-resolution spectrographs, even for the brightest stars.

The observational problem is further complicated by the timescales of pulsations, outflows and rotation, which are all of the order of hours to days, close to the 1-day rotation period of the Earth. Therefore, any serious attempt to study these phenomena requires the simultaneous effort of several observatories around the globe (*e.g.* the MuSiCoS 1998 campaign, see Neiner *et al.* 2002) or satellites.

Especially the detection of a beating effect between two close frequencies is very difficult, as it requires a precision in frequency of 0.01 c d^{-1}, *i.e.* many weeks of continuous observations. Yet, Rivinius *et al.* (1998) obtained the first evidence for pulsation beating, with spectroscopic observations of the Be star μ Cen. They could relate the beating period to epochs of strong emission events, and thus successfully predict the approximate time of future outbursts

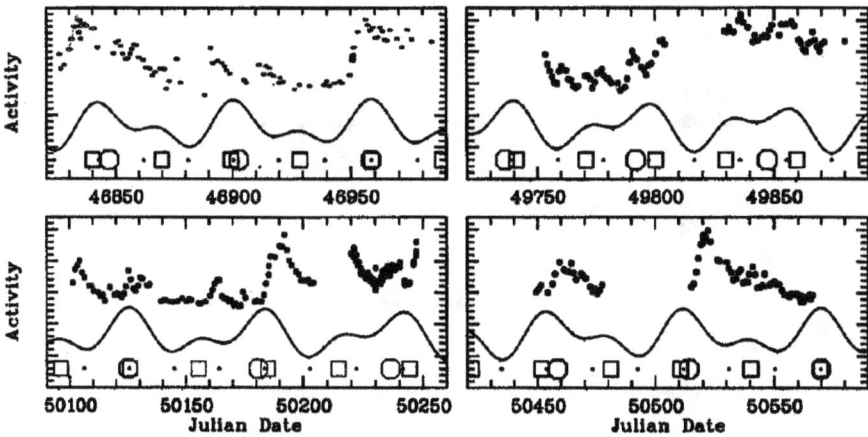

Fig. 16. Coincidence between the observed activity of μ Cen (top points) and the activity predicted from the beating effect of pulsations (solid line). The maximum amplitude of each pulsation mode is indicated at the bottom, with a different symbol for each pulsation mode. When the modes are at maximum at the same time (beating), ejection of material can occur. Taken from Rivinius *et al.* (1998).

(see Fig. 16). In 2009, Huat *et al.* (2009) observed an outburst in the Be star HD 49330 with the CoRoT satellite and simultaneous spectroscopic observations. These data allowed to detect several pulsation modes, whose amplitude vary with time. A clear correlation between the maximum amplitude of certain modes and the outburst was brought to light, thus demonstrating the direct link between pulsations and the formation of the circumstellar disk, at leats for some Be stars.

The CoRoT satellite also allowed to confirm that late Be stars host pulsations as well, by detecting pulsations with very low amplitude in these stars (*e.g.* Gutiérrez-Soto *et al.* 2009).

8   Magnetic field

It is well known that solar-type (low-mass) stars have magnetic fields, generated in the convective outer layer of the star. Massive stars, however, have only a very thin convective region near the surface (apart from the convective core) and no mechanism is known that can generate a substantial field in these stars. Nevertheless, a number of magnetic B stars are known! The presence of a magnetic field in Be stars would modify the energy balance and provide an explanation to the Be phenomenon. The Zeeman effect due to magnetic field, the magnetic oblique rotator model and observations of magnetic Be stars are presented in this section.

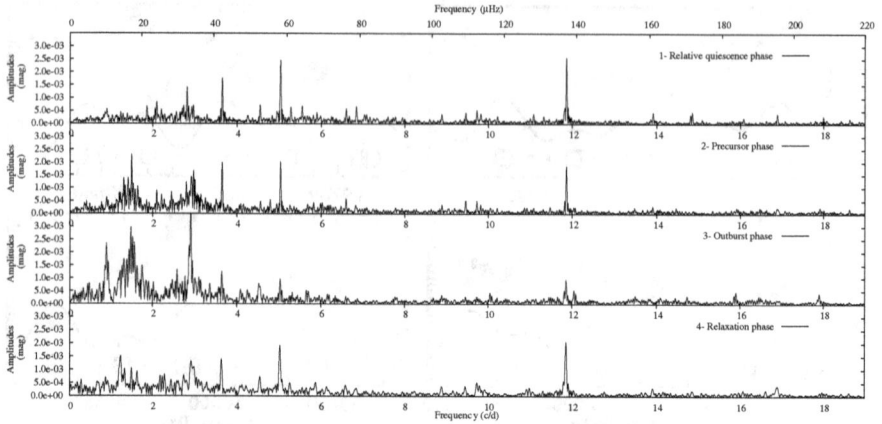

Fig. 17. Correlation between the amplitude of pulsations modes in the Be star HD 49330 and the various phases of the outburst (Huat *et al.* 2009).

8.1   Zeeman effect

When studying the formation of spectral lines in the presence of a magnetic field, it is necessary to take into account the state of polarisation of the light. The state of arbitrarily polarised radiation can generally be represented by a set of four parameters, first introduced by Stokes (1852). Chandrasekhar (1950) introduced the present form of the Stokes parameters: I, Q, U and V. I is the total intensity of the radiation, Q and U characterise the linearly polarised light, and V represents the circularly polarised light.

In the presence of a magnetic field, an atomic level is characterised by three fundamental quantities: its energy, the total angular momentum quantum number J and the Landé factor g. Because of the magnetic field, the atomic level is split into $(2J + 1)$ atomic states, each of them having a magnetic quantum number M which varies from $-J$ to J in steps of one. This splitting is called the "Zeeman effect" (also see Chapter 1).

Note that this description is only valid in the weak field limit, *i.e.* when the quadratic Zeeman effect is negligible and the magnetic splitting of the level is small compared to the fine structure, which is the case for Be stars. For more details and applications to astronomical settings, see Mathys (1989).

The presence of a magnetic field changes the energy balance of the star. In the case of a Be star, the magnetic field would bring the additionnal angular momentum needed to form the circumstellar disk. If the Be star is a slow rotator and has a strong enough dipolar magnetic field, the wind particles escaping from the star at the magnetic poles would follow the magnetic field lines towards the magnetic equator. The particles coming from both poles would then collide at the equator and a disk would be formed there. This is known as a magnetically confined disk. If the Be star is a rapid rotator, the magnetic field would have to be

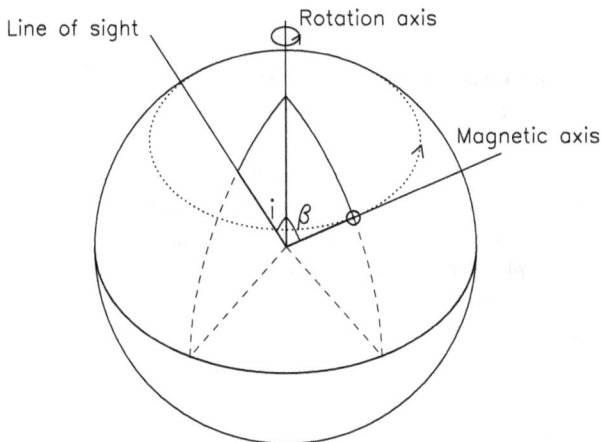

Fig. 18. Geometry of a rotating oblique magnetic dipole. The inclination angle i and the obliquity angle β are indicated. The dotted line shows the course of the magnetic pole over the stellar surface as the star rotates.

very strong to actually confine the disk. This is not the case of classical Be stars, where the rotation dominates. Thus the disk of classical Be stars is located at the stellar equator. However, the presence of a magnetic field makes the regions of intersection of the magnetic and stellar equators more stable. That is why clouds of denser material can form at the intersections of the two equators.

8.2 The magnetic oblique rotator model

In the magnetic oblique rotator model (Stibbs 1950) the magnetic field structure is not symmetric about the rotation axis of the star. For the most simple case of a dipole field, this means that the axis of the dipole and the axis of rotation of the star do not coincide. The observed stellar configuration can then be characterised by the inclination angle i between the observer's line-of-sight and the stellar rotation axis, and by the obliquity angle β between the magnetic axis and the rotation axis (see Fig. 18).

In the case of a dipolar magnetic star, the aspect of the visible hemisphere of the star changes as it rotates. This leads to variations in various observables, such as the shape and equivalent width of wind-sensitive UV resonance lines. The period of these variations is the stellar rotation period. It is therefore called "rotational modulation". The magnetic field also determines the properties of surface features, such as the distribution of the chemical abundances over the star. Thus, for a magnetic star, variations of the line profiles (*e.g.* the central depth of the lines) with the rotation period are also observed.

8.3 Observations

Over the past years an increasing number of observations has been obtained, which provides indirect evidence that hot stars must have magnetic fields, in particular because their winds are modulated or perturbed by the rotation. However, direct detection of weak magnetic fields in hot stars is particularly challenging. A high-resolution spectropolarimeter mounted on a large telescope is needed, and only very few such configurations exist. The present detection technique uses the difference in wavelength of the oppositely shifted Zeeman components of a spectral line as a measure of the strength of the magnetic field. For weak fields this wavelength difference is at the limit of what can be detected with present instrumentation.

The result can be improved by accumulating the magnetic signal of all the observed lines together, with the Least Square Deconvolution (LSD) technique (Donati et al. 1997). This technique has provided many detections of magnetic fields for cool stars, which have thousands of lines. However, the number of spectral lines in hot stars is much lower and their line profiles are often broadened by fast rotation and perturbated by other mechanisms (such as pulsations). Another complication is that the Zeeman technique only measures the projected component of the field in the line of sight, which can be much weaker than the polar field, especially when high order fields are involved.

Yet, Henrichs et al. (2000) have reported the detection of a weak magnetic field in β Cep, the prototype of β Cephei stars, which is also a binary star with a Be companion (Fig. 19). The magnetic field of this star is well understood in the frame of an oblique rotator model with a magnetically confined disk.

Recently, strong evidence for the presence of a magnetic field in a classical Be star was for the first time detected (Neiner et al. 2003). The star ω Ori seems to host a magnetic dipole field of about 500 G at the poles, with an axis inclined by 50°. The longitudinal magnetic field varies with the rotation period of about 1.3 d, as well as the UV highly ionised resonance lines sensitive to the wind. The disk of this star is at the stellar equator, but clouds have been detected at the intersections of the stellar and magnetic equators (see Fig. 20). These promising evidences of the presence of a magnetic field have unfortunately not been confirmed by the clear detection of Zeeman signature detection in the Stokes V profiles in further observations. The possibility of a transiting character of the magnetic field of Be stars should thus be considered.

9   Spectroscopy of Be stars by amateurs

If Be stars remain a mystery for professional astronomers since their discovery almost one and a half century ago, they also fascinate amateurs. Indeed the observation of a Be star always provides a new surprise for the observer, even if the same Be star has already been observed before. Neither two Be stars nor two spectra of a given Be star are identical.

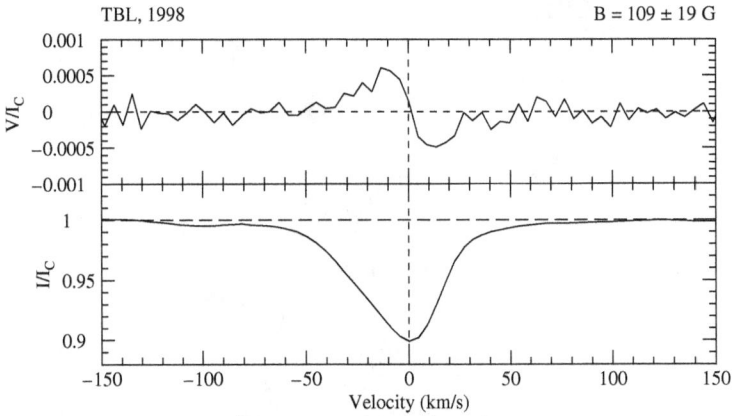

Fig. 19. Magnetic field measurement of the star β Cep, which was taken close to the phase of maximum field. β Cep is a binary star with a Be secondary, but it is the primary B star which is magnetic. The top panel shows the Stokes V profile, *i.e.* the Zeeman signature of circularly polarised light. The bottom panel shows the Stokes I profile, *i.e.* the mean photospheric line, which is skewed to the left because of the pulsation phase.

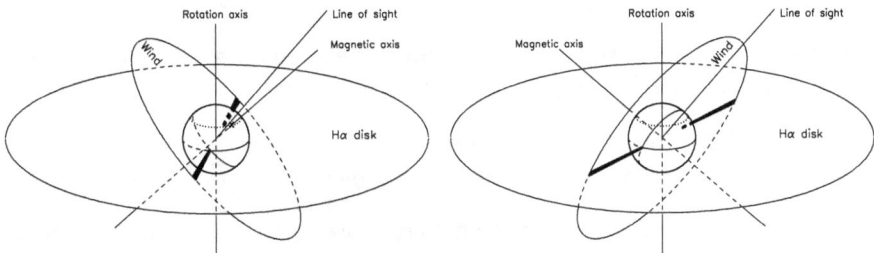

Fig. 20. Geometrical configuration of ω Ori. The magnetic axis is not aligned with the rotation axis, and thus the magnetic and rotation equators do not coincide. The disk is at the stellar equator, but at the intersections of the two equators are two clouds of denser material.

9.1 Small wavelength domain, low resolution: Hα

The French amateurs Alain Klotz (now a professional astronomer), Valérie Desnoux and Daniel Barbin started a survey of the Hα line of Be stars in 1992, with a 60 cm telescope at the Pic du Midi Observatory (France). This survey was taken over by Christian Buil and his Takahashi CN-212 equipped with a spectrograph with a resolution of 7000 covering 600 Å and a Audine CCD camera. About 300 Be stars are regularly monitored. The data are available online at http://www.astrosurf.com/buil/us/bestar.htm. Figure 21 shows observations obtained by C. Buil during an outburst of δ Sco.

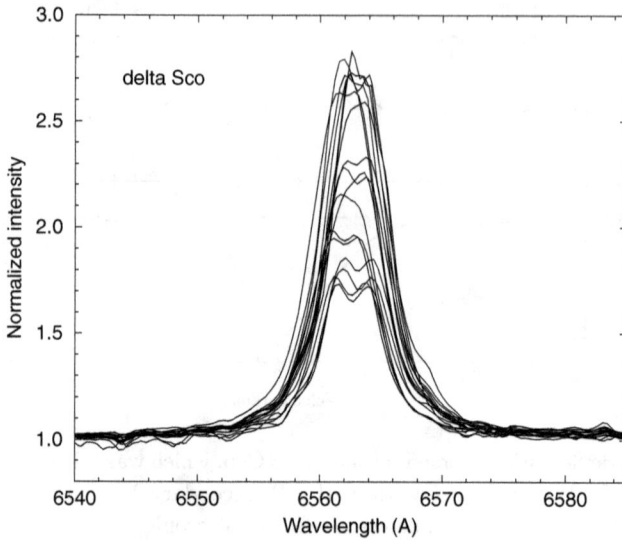

Fig. 21. Observation of an outburst in the Be star δ Sco, by the amateur Christian Buil.

Such a survey of the Hα line profile can be easily done with a low-resolution spectrograph on a small telescope but is of high interest:

- It allows us to follow the long-term evolution of the emission in Hα. We can then know the timescale of variation, the recurrence of outbursts, etc.

- It allows us to know rapidly when a star enters an outburst phase and thus to follow the evolution of the outburst.

- It allows us to study the long-term V/R variations in Hα and thus the global disk oscillation.

- If the survey also concentrate on normal B stars, it allows us to compare the behaviour of B and Be stars, as well as to discover new Be stars.

9.2   Longer wavelength domain, higher resolution

Be stars are also one of the focus of ARAS (Astronomical Ring for Access to Spectroscopy, http://astrosurf.com/aras/), a collaboration between amateurs and professionals. In particular, in September 2003, a team of amateurs led by Olivier Thizy installed a copy of the high resolution (R = 35 000) Musicos spectrograph (kindly put at their disposal by ESA) on the 60 cm telescope of Saint Véran (France), which is administrated by the amateur association Astroqueyras (http://www.astroqueyras.com). They obtained the first amateur high resolution échelle spectra of Be stars.

Of course, observations of the Hα line profile at higher resolution provides more informations than at low resolution:

- It allows us to study the short-term variations of the line-profile due to pulsations and rotational modulation, including the short-term variations of the V/R ratio, equivalent widths, etc.

- It allows us to detect transient phenomena, such as the ejection of a cloud of material which then dilutes in the disk

Observations on a wider wavelength domain also provide additionnal informations:

- The comparison of the line-profile variations of different lines makes the determination of pulsation and rotation properties easier.

- The comparison of the level of emission in different Balmer lines gives us an indication of the size of the disk.

- Other emission lines can be studied, *e.g.* HeI lines.

9.3   BeSS et ArasBeam

The creation of the professional-amateur BeSS database (`http://basebe.obspm.fr`) at the Paris-Meudon Observatory and of the ArasBeam tool for the planning of observations (`http://arasmean.free.fr`) lead by François Cochard allowed to organize amateur observations on a greater scale.

BeSS contains a catalog of all known Be stars and allows anyone who wishes so to deposit obtained Be star spectra in a database, which astronomers can then use for scientific studies. ArasBeam allows to coordinate these observations among all interested amateurs by attributing observing priorities for the coming nights for the various Be stars. Moreover, the spreading of very efficient spectrographs at relatively low cost by the Shelyak Instruments company (`http://www.shelyak.com`) allowed to make this king of programs popular and to homogenize the nature and quality of observations obtained by amateur astronomers.

9.4   Reduction of spectra: Point of caution

When observing Be stars, apart from the usual reduction process of the stellar spectrum (see Chapter 3), one should be very careful when determining the continuum of the spectrum. The normalisation process is indeed crucial, not only for determining the emission strength but especially for the study of line-profile variations. A badly fitted continuum level would ruin the analysis. That is why spectra inserted in the BeSS database are usually not normalized, thus allowing each scientist to perform the normalisation him/herself depending on the needs of his/her study.

10   Conclusions

Be stars show line emission associated with the presence of a circumstellar disk. How this disk is created is not understood yet, but rotation, pulsations and magnetic fields are probably part of the explanation. Spectroscopy is thus a useful tool to study line-profile variations related to these physical processes and the parameters of Be stars.

Professional astronomers and amateurs work together to try to understand the Be phenomenon. In this way they can reach the necessary time sampling and wavelength coverage to study the different timescales of variation of many Be stars and hope to solve this mystery.

References

Andrillat, A., Jaschek, M., & Jaschek, C., 1990, A&AS, 84, 11

Andrillat, Y., & Fehrenbach, C., 1982, A&AS, 48, 93

Baade, D., 2000, in IAU Coll. 175, The Be Phenomenon in Early-Type Stars, ed. M.A. Smith, H.F. Henrichs & J. Fabregat, ASP Conf. Ser., 214, 178

Balona, L.A., 2000, in IAU Coll. 175, The Be Phenomenon in Early-Type Stars, ed. M.A. Smith, H.F. Henrichs & J. Fabregat, ASP Conf. Ser., 214, 1

Balona, L.A., Aerts, C., Božić, H., et al., 2001, MNRAS, 327, 1288, B01

Banerjee, D.P.K., Rawat, S.D., & Janardhan, P., 2000, A&AS, 147, 229

Bopp, B.W., & Dempsey, R.C., 1989, Inf. Bull. Variable Stars, 3387, 1

Briot, D., 1977, A&A, 54, 599

Buil, C., 2001, Atlas of Be stars on the web [http://www.astrosurf.com/buil/us/bestar.htm]

Chandrasekhar, S., 1950, Radiative transfer (Oxford, Clarendon Press, 1950)

Chauville, J., Zorec, J., Ballereau, D., et al., 2001, A&A, 378, 861

Cohen, D.H., Cassinelli, J.P., & Macfarlane, J.J., 1997, ApJ, 487, 867

Collins, G.W., 1987, in IAU Colloq. 92, Physics of Be Stars, 3

Dachs, J., Eichendorf, W., Schleicher, H., et al., 1981, A&AS, 43, 427

Divan, L., 1979, in IAU Colloq. 47, Spectral Classification of the Future, 247

Doazan, V., Sedmak, G., Barylak, M., & Rusconi, L., 1991, ESA-SP 1147

Domiciano de Souza, A., Kervella, P., Jankov, S., et al., 2003, A&A, 407, L47

Donati, J.-F., Semel, M., Carter, B.D., Rees, D.E., & Collier Cameron, A., 1997, MNRAS, 291, 658

Floquet, M., Hubert, A.M., Hirata, R., et al., 2000, A&A, 362, 1020

Floquet, M., Neiner, C., Janot-Pacheco, E., et al., 2002, A&A, 394, 137

Frémat, Y., Zorec, J., Hubert, A., & Floquet, M., 2005, A&A, 440, 305

Grady, C.A., Bjorkman, K.S., Snow, T.P., et al., 1989, ApJ, 339, 403

Gutiérrez-Soto, J., Floquet, M., Samadi, R., et al., 2009, A&A, 506, 133

Hanuschik, R.W., Hummel, W., Sutorius, E., Dietle, O., & Thimm, G., 1996, A&AS, 116, 309

Henrichs, H.F., de Jong, J.A., Donati, J.-F., *et al.*, 2000, in IAU Coll. 175, The Be Phenomenon in Early-Type Stars, ed. M.A. Smith, H.F. Henrichs & J. Fabregat, ASP Conf. Ser., 214, 324

Hirata, R., & Hubert-Delplace, A.M., 1981, in Pulsating B-Stars, 217

Huat, A., Hubert, A., Baudin, F., *et al.*, 2009, A&A, 506, 95

Hubert, A.M., & Floquet, M., 1998, A&A, 335, 565

Koubský, P., Harmanec, P., Hubert, A.M., *et al.*, 2000, A&A, 356, 913

Mathys, G., 1989, Fund. Cosmic Phys., 13, 143

Neiner, C., Hubert, A.-M., Floquet, M., *et al.*, 2002, A&A, 388, 899

Neiner, C., Hubert, A.-M., Frémat, Y., *et al.*, 2003, A&A, 409, 275

Oudmaijer, R.D. & Drew, J.E., 1999, MNRAS, 305, 166

Poeckert, R., Bastien, P., & Landstreet, J.D., 1979, AJ, 84, 812

Quirrenbach, A., Bjorkman, K.S., Bjorkman, J.E., *et al.*, 1997, ApJ, 479, 477

Rivinius, T., Baade, D., Stefl, S., *et al.*, 1998, A&A, 336, 177

Schrijvers, C., 1999, Ph.D. Thesis, Universiteit van Amsterdam

Secchi, A., 1867, Astron. Nachr., 68, 63

Slettebak, A., 1979, Space Science Rev., 23, 541

Sonneborn, G., Grady, C.A., Wu, C., *et al.*, 1988, ApJ, 325, 784

Srinivasan, K., 1996, Ph.D. Thesis, Bharathidasan University

Stee, P., 2000, in IAU Coll. 175, The Be Phenomenon in Early-Type Stars, ed. M.A. Smith, H.F. Henrichs & J. Fabregat, ASP Conf. Ser., 214, 129

Steele, I.A., Negueruela, I., & Clark, J.S., 1999, A&AS, 137, 147

Stibbs, D.W.N., 1950, MNRAS, 110, 395

Stokes, G.G., 1852, Trans. Cambridge Phil. Soc., 9, 399

Struve, O., 1931, ApJ, 73, 94

Telting, J., 1997, Ph.D. Thesis, Universiteit van Amsterdam

Townsend, R.H.D., Owocki, S.P., & Howarth, I.D., 2004, MNRAS, 350, 189

Unno, W., Osaki, Y., Ando, H., Saio, H., & Shibahashi, H., 1989, Nonradial oscillations of stars, 2nd ed. (University of Tokyo Press)

Vakili, F., Mourard, D., Stee, P., *et al.*, 1998, A&A, 335, 261

Vogt, S.S., & Penrod, G.D., 1983, ApJ, 275, 661

Waters, L.B.F., Marlborough, J.M., van der Veen, W.E.C., Taylor, A.R., & Dougherty, S.M., 1991, A&A, 244, 120

Astronomical Spectrography for Amateurs
J.-P. Rozelot and C. Neiner (eds)
EAS Publications Series, **47** (2011) 165–188

COMETARY SPECTROSCOPY

N. Biver[1]

Abstract. Cometary spectroscopy from the ultraviolet to the radio wavelength domain provides us with insights on the composition of the gases that are released by the cometary nuclei. While infrared to millimeter spectroscopy give access to the parent molecules that are released directly from the nucleus, visible spectroscopy enables observation of daughter species. Those "radicals" observable in the visible domain have more complex spectroscopic band-like structures and are mainly CN, C_2, C_3, NH_2. Their spectroscopic signatures are easily accessible to amateur astronomers class equipment. Provided that carefully calibrated data are acquired, some simple calculation can readily be done to convert the line intensities into comet molecular outgassing rates and thus provide interesting physical data on comets. In addition to broadband dust measurements, the interested amateur can produce valuable scientific data on comets that will always be welcome from the professional community and certainly useful as the monitoring of comets activity is always essential.

1   Introduction

Spectroscopic study of comets provides information on the physical and chemical characteristics of the coma surrounding their nucleus. Molecules sublimating from the nucleus ("parent molecules") and their photo-dissociation products have narrow spectral signatures in the ultraviolet to visible, infrared and radio wavelengths domains, depending on the emission mechanism. Very high spectral resolution ($\lambda/\Delta\lambda \geq 10^6$) is necessary to get information on the gas expansion velocity. Only radio techniques can provide it (Figs. 6–8). On the other hand, wide band spectroscopy can show the emission spectrum of dust in the mid- to far-infrared (5–$100\,\mu$m) domain (Figs. 1, 3) and the scattered sun light in the visible range (Figs. 1, 2).

[1] LESIA, UMR8109 du CNRS, Observatoire de Paris-Meudon, 5 place Jules Janssen, 92195 Meudon Cedex, France; e-mail: `nicolas.biver@obspm.fr`

© EAS, EDP Sciences 2011
DOI: 10.1051/eas/1147006

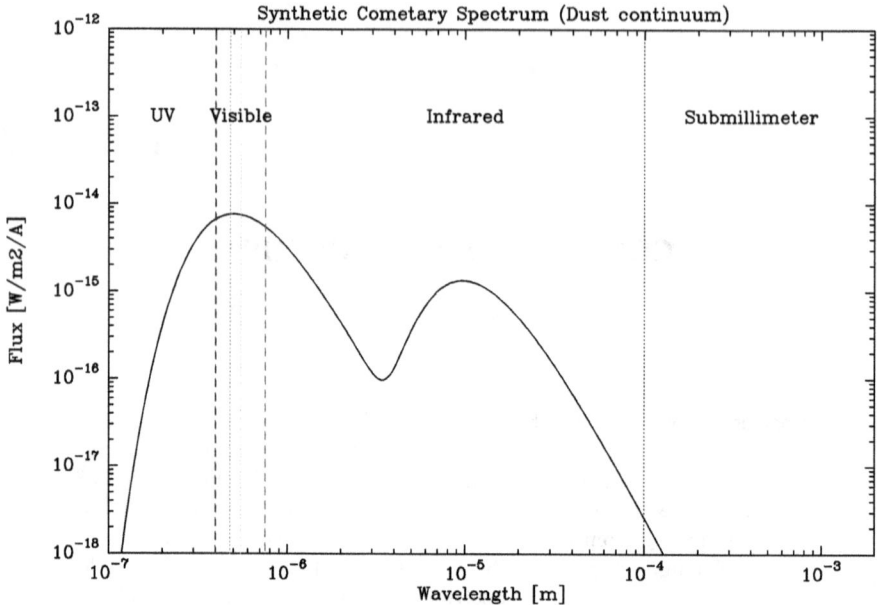

Fig. 1. Very wide band (UV to radio) synthetic spectrum of a bright comet.

In the ultraviolet-visible domain, cometary spectra are dominated by spectral lines of radicals, unstable molecules which come from parent molecules which have been stripped of one or a few atoms by the solar radiation. The main radicals are C_2, responsible for the famous cometary "Swan bands" and the green tint of cometary comae, CN (violet), C_3, NH, NH_2 and OH (Fig. 2). Line intensities are used to determine the amount of molecules in the coma. On the other hand, the integrated intensity of the continuum can be used to estimate the quantity of dust in the coma but strongly depends on its physical characteristics: size distribution of grains, scattering properties,...

We will start with an inventory of comet molecules and investigation techniques before focusing on the analysis of visible comet spectra which is a discipline accessible to amateur astronomers.

2   The observation of cometary molecules

The main specificity of cometary nuclei is to be composed of a large fraction of ices of volatiles such as water but also CO and CO_2, for the main components. As the comet nucleus gets close to the sun its ices sublimate to create the cometary atmosphere (coma) and tails of ionized gas and dust. Dust comes out from the nucleus leveled off by the gas.

Although water has been suspected as the major component of cometary ices for over a century, it was only directly observed for the first time in 1986. However,

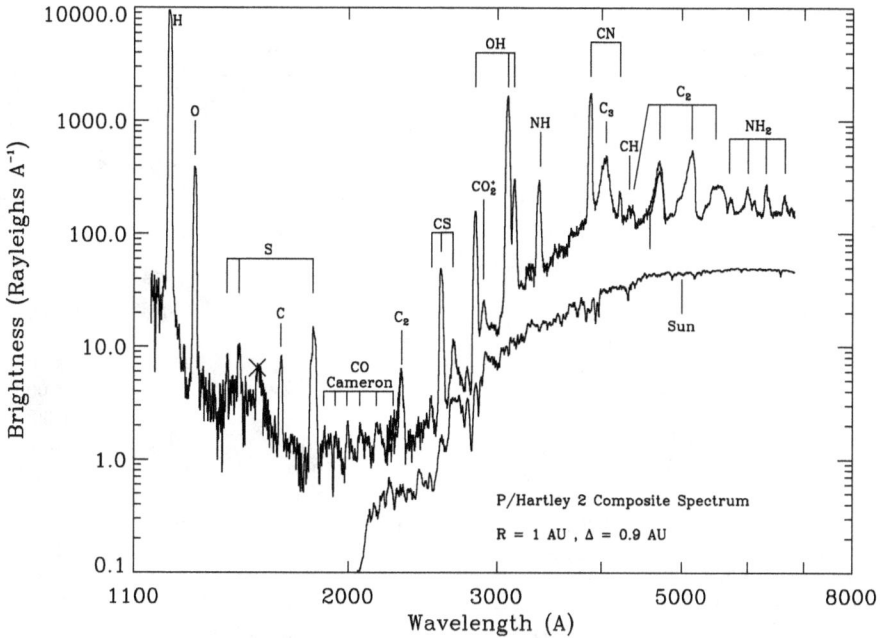

Fig. 2. Ultraviolet to visible spectrum of comet 103P/Hartley-2 observed with the Hubble Space Telescope (Weaver & Feldman 1992; Festou *et al.* 1993). In addition to the classical visible radical and ion bands on the right, in the UV one can notice atomic lines and the strong H Lyman-α line at 1216 Å. Atomic hydrogen, mostly coming from the photo-dissociation of water, is the most abundant species in cometary comae.

the visible signatures (Fig. 2), C_2 green SWAN bands, CN cyanogen line (The presence of cyanogen in comets was responsible for the 1910 panic at the time the earth crossed Halley's comet tail) and atomic lines seen in sun-grazing comets have been identified for over a century. But these only trace decomposition products of the main "parent" molecules coming directly from the nuclear ices sublimation. Radio and infrared techniques are the ones responsible for the identification of over 22 parent cometary molecules between 1985 and 1997 (Table 1).

Figures 2 and 3 give an overview of typical comet spectra with both continuum and spectral lines. After the inventory of cometary lines, we will look at the different techniques and frequency ranges to observe these cometary molecules.

2.1 The observed cometary molecules

Table 1 hereafter provides the list of the majority of molecules and radicals observed in comets, their mean relative abundance to water, and parent scale-length (L_p) for a photo-dissociation product and dissociation scale-length (L_d) that will be useful to evaluate their production rate and abundance (Sect. 4). Scale-lengths are given at an heliocentric distance r_h of 1 AU, and often scale as r_h^2. Several

Table 1. Observed cometary molecules.

Molecule	Radio	Main lines Infrared	Visible-UV	Abundance [% water]	L_p [km]^d	L_d [km]^d
H_2O	0.54–0.17 mm[1]	2.7, 6.3 μm^a, 1.94 2.95 μm 4.65 μm	–	100	0	70 000
OH	18 cm	2.87 μm^a 3.04, 3.28 μm	0.30 μm	90	25 000	160 000
H	–	–	121.6 nm^a	200	$\approx 10^5$	3 10^7
CO_2	–	4.25 μm^a	(from CO^b:) (185–230 m^a)	5–10	0	430 000
CO	2.6–0.65 mm	4.67 μm	142–160 nm^a	1–25	0+	1.3 10^6
CH_4	–	3.31 μm	–	0.2–0.8	0	105 000
C_2H_2	–	3.03 μm	–	0.3	0	60 000
C_2H_6	–	3.35 μm	–	0.1–0.7	0	75 000
C_2	–	–	0.45–0.56 μm	0.01–0.70	20 000	70 000
C_3	–	–	0.405 μm	0.003–0.07	2500	20 000
CH	–	3.35 μm	0.431 μm	0.05–0.5	80 000	5000
CH_3OH	3–0.6 mm	3.52 μm	–	0.5–6	0	60 000
H_2CO	2.1–0.8 mm	3.59 μm	–	0.1–1.2	7500	5000
HCOOH	1.3 mm	–	–	0.09	0	27 000
CH_3CHO	2–1 mm	–	–	0.02	0	12 000
$HCOOCH_3$	1.3 mm	–	–	0.08	0	18 000
$(CH_2OH)_2$	3–1 mm	–	–	0.25	0	10^5?
NH_3	1.3 cm	3.00 μm	–	0.7	0	5500
NH_2	–	3.23 μm	0.52–0.74 μm	0.2	5000	10 000
NH	–	–	0.336 μm	0.3	50 000	150 000
HCN	3.4–0.4 mm	3.0 μm	–	0.08–0.25	0	57 000
CN	1.3 mm	4.90 μm	0.388 μm	0.1–0.6	20 000	200 000
HNC	3.3–0.8 mm	–	–	0.005–0.02	0 ?	57 000
HNCO	1.4–0.9 mm	–	–	0.1	0	29 000
CH_3CN	3.3–1.3 mm	–	–	0.01	0	110 000
HC_3N	3.3–1.1 mm	–	–	0.01	0	13 000
NH_2CHO	1.3 mm	–	–	0.01	0	10^4?
H_2S	1.8, 1.4 mm	–	–	0.4–1.5	0	4000
OCS	2.0–1.0 mm	4.86 μm	–	0.4	0+	9000
CS (CS_2)	3.1–0.9 mm	–	260 nm	0.1	300	40 000?
SO_2	1.5–1.3 mm	–	–	0.2	0	4000
SO	1.4–1.0 mm	–	–	0.3	4000?	6000
H_2CS	1.3 mm	–	–	0.02	0	?000
NS	0.9 mm	–	–	0.02	?	?
S_2	–	–	290 nm	0.005	0 ?	200
H_2O+		3.1 μm	550–747 nm	0.2%($r_n = 10^5$)–2%($r_n = 10^6$ km)		
H_3O+	1.0 mm	2.8 μm	–	0.01% maxi at $r_n = 10^4$ km^c		
$CO+$	1.3 mm	–	340–630 nm	0.1%($r_n = 10^5$)–30% ($r_n = 10^6$ km)		

^{a} Non observable from the ground;

^{b} Photo-dissociation product of CO_2 in an excited state;

^{c} r_n = distance from comet nucleus in km;

^{d} Using $v \approx 0.8$ km s^{-1} for life-time to scale-length conversion when possible.

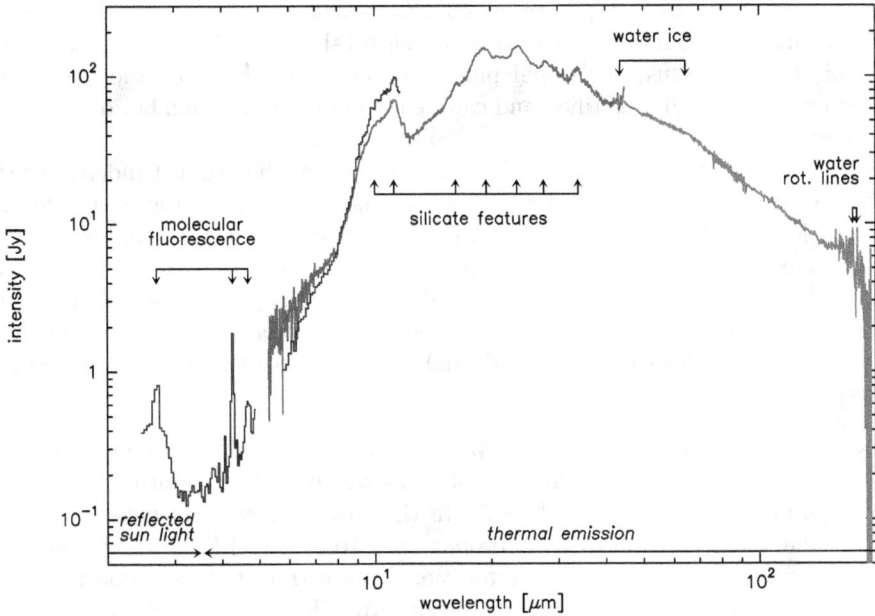

Fig. 3. Near to far infrared spectrum of comet C/1995 O1 (Hale-Bopp) observed with the Infrared Space Observatory Crovisier 2000. In addition to the blackbody spectrum of the dust, several silicate emission bands appear in the middle of the spectrum. On the other hand a few molecular lines are seen on each side (*left*: vibrational bands, *right*: rotational lines) of the spectrum.

radicals are photo-dissociation products of well known parent molecules, such as:

- $H_2O \rightarrow OH, H$;

- $HCN \rightarrow CN$;

- $C_2H_2, C_2H_6 \rightarrow C_2$;

- $NH_3 \rightarrow NH_2, NH$.

3 Comets molecular spectroscopy

3.1 Short introduction to molecular spectroscopy

The purpose of this section is not to make a detail presentation of the principles of molecular spectroscopy, but just give some simple basics and examples to understand the difference between different wavelength spectra. Generally the wave function describing a molecule state is made of 3 main components: electronic function, vibrational function and rotational function. The last two functions are not relevant to isolated atoms. In most cases (because of high energy differences) those 3 functions can be decoupled, and the total molecular energy will be the sum of electronic plus vibrational plus rotational energies (from the highest to lowest), which are all quantified and can be estimated from a number of quantum numbers.

To the first approximation each energy mode can be studied independently and the coupling between modes will not change significantly the energy levels. We will look at the case of the CO molecule, abundant in comets and one of the simplest example. Generally the complexity and lack of symmetry of a molecule will make the spectra more complex and require a larger number of quantum numbers to describe the energy states. They can be even further more split into energy sub-levels (especially for radicals) and lead to more complex hyperfine spectral structures.

- Rotational states can be described by 1 to 3 (general case) quantum numbers (*e.g.* J, Ka, Kb): linear molecules will need one quantum number J, symmetric ones 2, and others 3. In the case of linear molecules, the rotational energy is to first order proportional to $J(J+1)$ (Fig. 4) and thus the frequency of the $J \rightarrow J - 1$ rotational state transition is proportional to $2 \times J$. Only the $\Delta J = 0, \pm 1$ transitions are allowed.

- Vibrational states: the more atoms the molecule has, the larger the number of vibrational mode it will have: *e.g.* only one for CO (C–O binding elongation), 3 for CO_2 or H_2O, 12 for CH_3OH.... They are generally called ν_1, ν_2, ... and to first approximation the energy level in each vibration mode is proportional to the v quantum number $+1/2$, thus $v = 2 \rightarrow 1$ and $v = 1 \rightarrow 0$ transition will correspond to very similar energy changes and result in spectroscopic lines at close frequencies. But all these vibrational levels have a rotational fine structure and we generally talk about "vibrational bands" since within a given Δv transition (all allowed) there are many rotational transitions possible. For linear molecules, the $\Delta J = 0, \pm 1$ selection rule results in three different series of lines "P ($\Delta J = +1$), Q($\Delta J = 0$), R ($\Delta J = -1$) branches" – for CO the Q branch is forbidden.

- Electronic states: labeling of the molecular electronic states is done in a similar way to atomic electronic states: they mostly concern energy levels of the outer electron(s). Energies are again much higher than for rotational and vibrational states, but there are often only a few electronic states of interest since other are generally dissociative for the molecule (energy higher than dissociation energy of the molecule). Figure 5 shows an example of the main two series of electronic levels for CO (with transitions), with their rotational and vibrational fine structure. Vibrational structure has been enlarged for clarity, with respect to the vertical energy scale. This can result in a dense

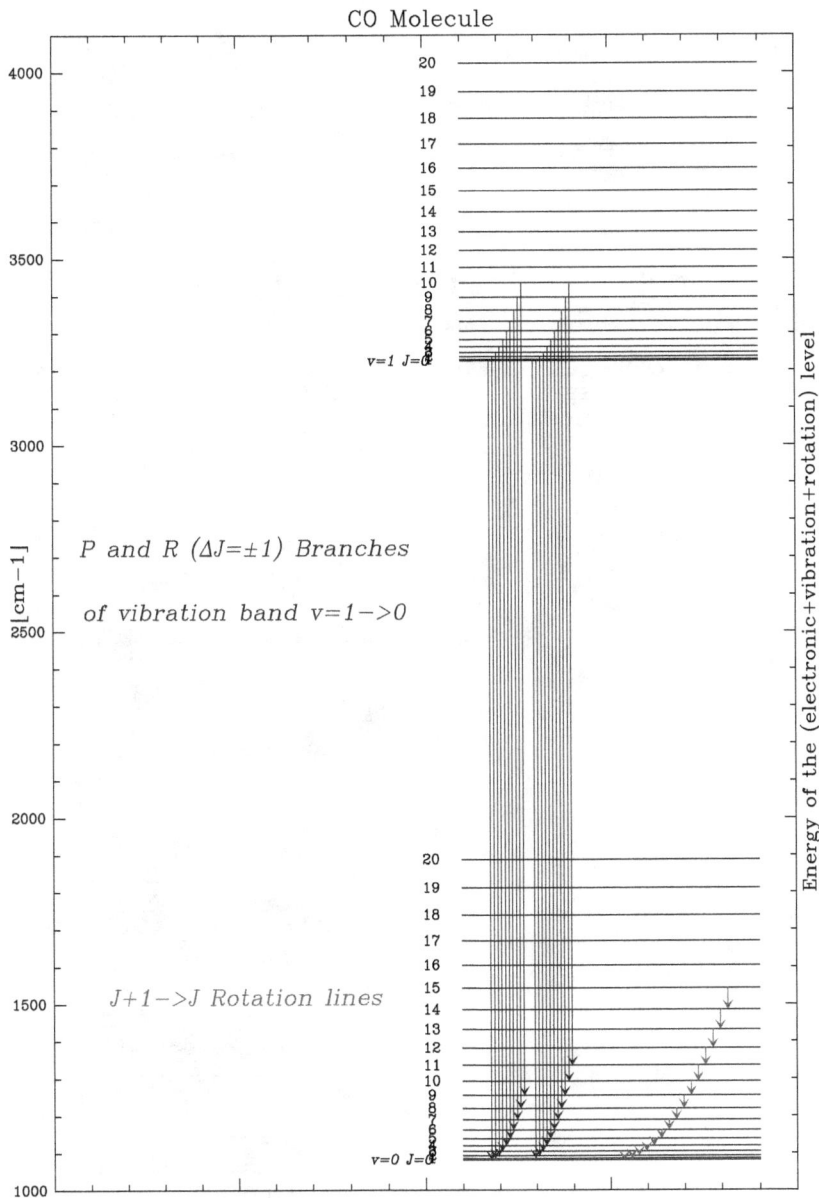

Fig. 4. Rotation states and rotational transition (at right) and first vibration band of CO with its rotational structure an corresponding transitions. Vertical scale: energy level in cm^{-1}.

forest of line at high spectral resolution, like in the case of C_2 (Fig. 10) where they are grouped by $\Delta v = 1, 0, -1, -2$ for the vibrational structure transition within the electronic transition – see Section 3.4 too.

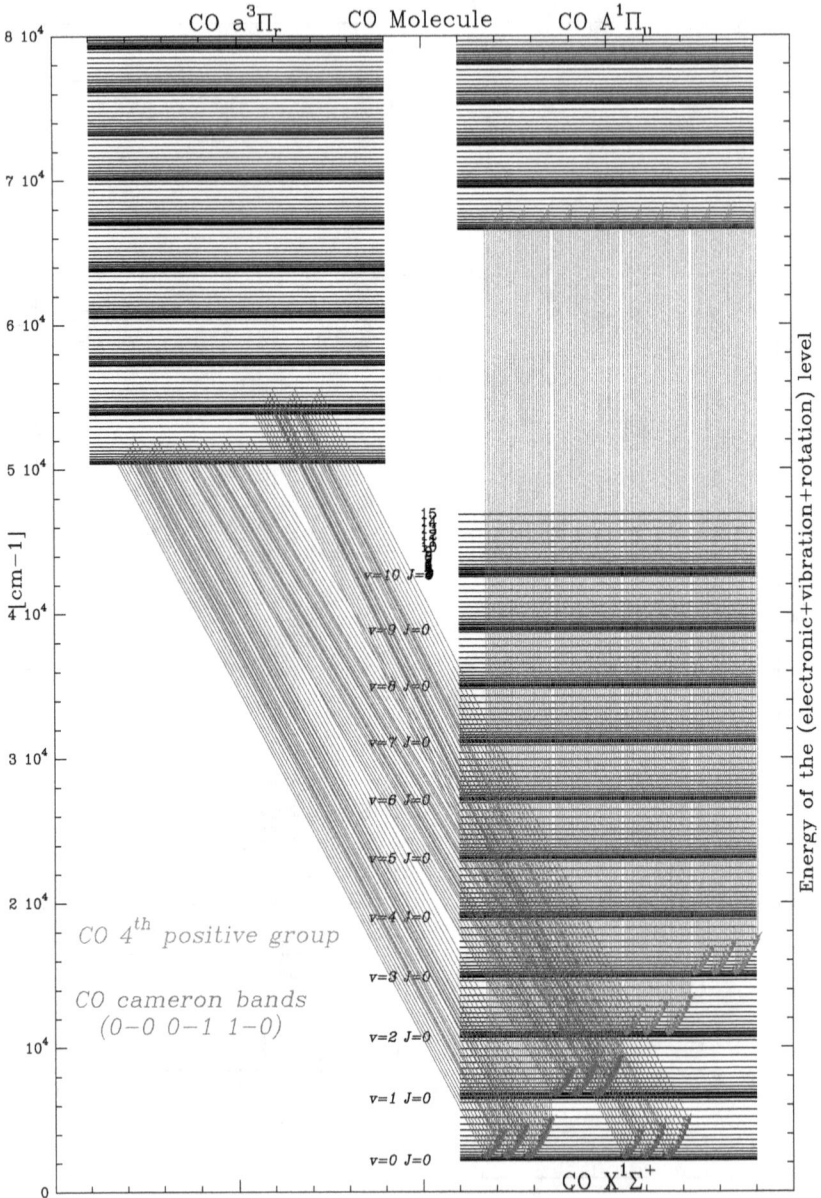

Fig. 5. UV electronic bands of CO.

3.2 Rotational spectra of molecules

- Spectral domain: radio sub-millimetric to centimetric ($\lambda \approx 0.1$–$10\,\mathrm{mm}$).

- Observable molecules: all that have a permanent dipole momentum because of their asymmetry (*e.g.* HCN but not CO_2, nor CH_4).

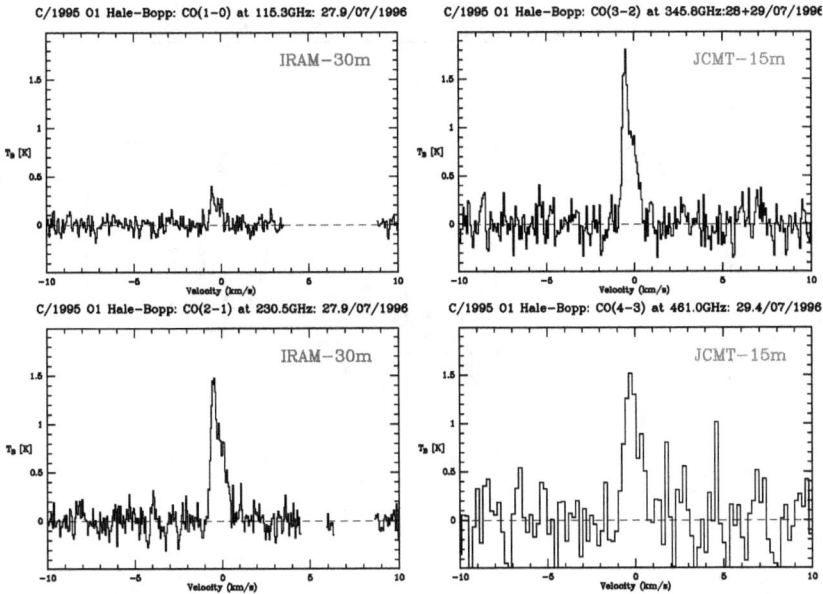

Fig. 6. Radio spectra of the 4 first transitions of CO observed in comet C/1995 O1 (Hale-Bopp) in 1996. The temperature inferred from the relative intensities of the lines was 30 K (Biver 1996). Transitions are shown in Figure 4.

The interest of this wavelength range is to relatively easily detect parent molecules: the cometary atmosphere is very cold (gas temperatures are in the 10–150 K range) and such lines correspond to transitions between low energy molecular levels. The observation of groups of lines such as those from methanol can probe the gas temperature.

The (radio) heterodyne technique offers access to ultra high spectral resolution (up to 10^8 currently at the Institut de Radioastronomie Millimetrique 30 m facility). It can be used to resolve the lines in Doppler velocity. Cometary lines are very narrow ($\Delta\lambda/\lambda \approx 10^{-5}$), since only broadened by the velocity dispersion due to the outgassing geometry. The gas is cold and of very low density and has a mean radial expansion velocity of 0.5 to 1.5 km s^{-1}. See example of Doppler-velocity resolved lines in Figures 6–8 and interpretation.

Lines are usually simple and well separated from each other and no confusion is possible. Over 200 lines have been identified in comets and none is still waiting for identification. There may be a few exceptions but these are marginal features: very weak lines necessitating hours of integration on bright comets with large telescopes (10–30 m diameter) at high altitude sites – well beyond amateur capacities.

Usually radio intensity (I) units are converted into temperatures from: $T = \lambda^2/(2k)I$ [Kelvin]. T is the equivalent brightness temperature of the black body that would radiate I (in the Rayleigh-Jeans approximation $\lambda T \gg 3000 \, \text{K} \, \mu\text{m}$).

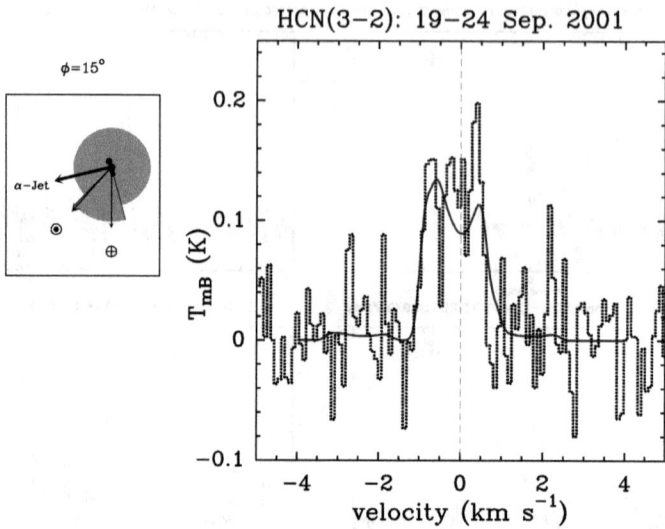

Fig. 7. Radio spectrum of the HCN(3-2) line at 1.1 mm in comet 19P/Borrelly in September 2001 observed with the IRAM 30 m radio-telescope (doted line). On the left: modeling of outgassing pattern yielding a line profile compatible with observations (Bockelée-Morvan *et al.* 2004).

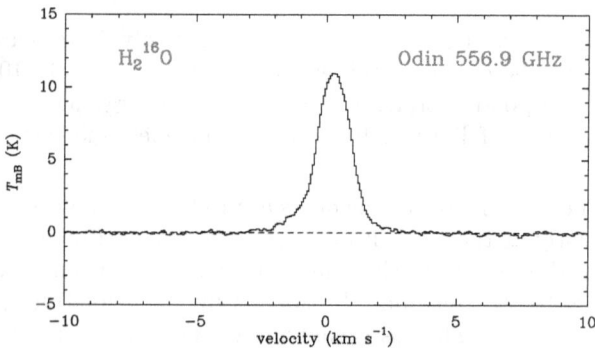

Fig. 8. Spectrum of water at 0.5 mm in comet 153P/Ikeya-Zhang from space with the ODIN satellite (1.1 m sub-millimeter radio-telescope) (Lecacheux *et al.* 2003). Here, asymmetry is due to self-absorption in an optically thick line.

3.3 Vibrational spectra of molecules

- Spectral domain: infrared ($\lambda \approx 2 - 10\,\mu m$), requiring dry high altitude site;

- Observable molecules: all excepted homo-nuclear molecules such as O_2, N_2 or S_2.

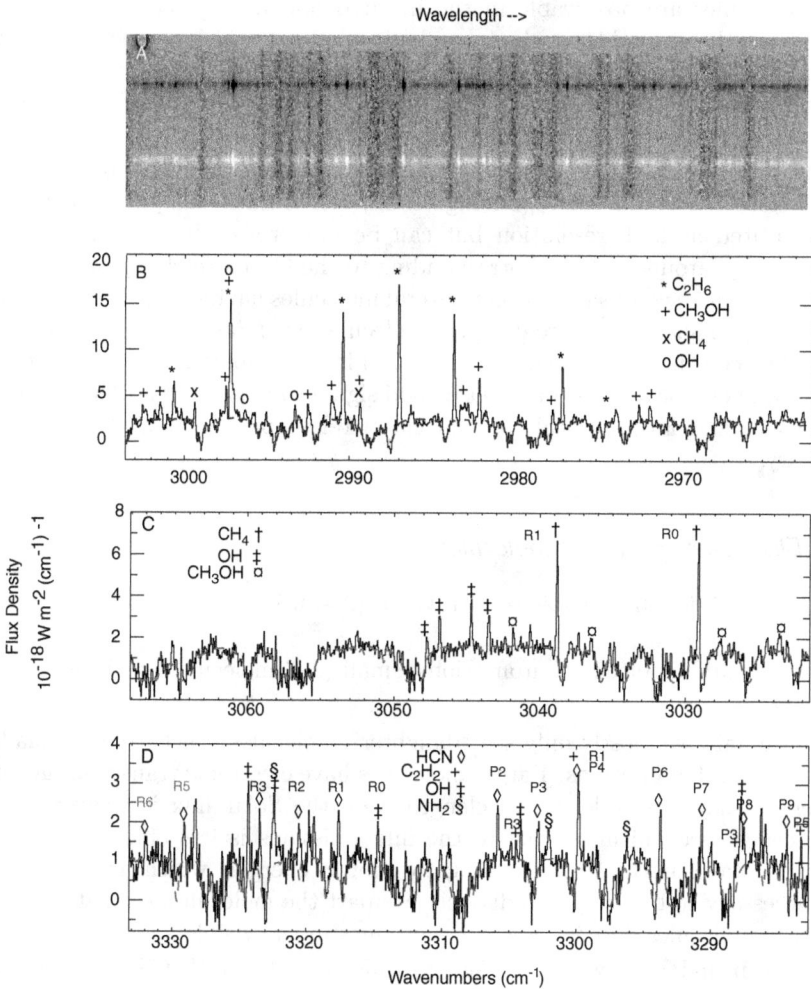

Fig. 9. Infrared spectra of comet C/1999 H1 (Lee) observed with the Keck telescope (and Echelle spectrometer NIRSPEC) (Mumma *et al.* 2001). The top plot (A) gives the spectro-image in order 23 with an horizontal cut (B spectrum) below. To cancel atmospheric background, the comet photocenter is moved along the slit from position 1 to 2 (12″ above) and the signal subtracted from the previous integration. The full integration results from the "1–2–2 + 1" series of integration/subtraction to cancel as far as possible atmospheric signal and fluctuations and is repeated as much as necessary to get a good detection. The final result is a positive (white) and negative (black) (A) spectro-image from which spectra are extracted. We can note noisy vertical bands in A corresponding to zero level signal in B, where the atmosphere is opaque and fully absorbs the continuum. An atmospheric transmission profile is overploted in dashed lines on each spectra. C and D correspond to other frequency ranges observed simultaneously in other dispersion orders. The spectral resolution is around 25 000.

Most molecules are observable in the infrared and a large number can be observed from the ground through atmospheric windows, which namely exclude CO_2 and most of H_2O lines because of telluric absorption. This technique brings the possibility of detecting symmetrical molecules (such as hydrocarbons like CH_4, C_2H_2, ...) that have no radio signatures.

The ro-vibrational lines (for a given vibrational transition there is a large number of close transition corresponding to different rotational quantum numbers) are separated at high resolution but can be numerous. In certain wavelength domains, e.g. around $3.4\,\mu m$, corresponding to the C–H stretching mode of vibration, there can be confusion between several molecules having very close transitions (see Fig. 9). Thus some more complex molecules are more easily identified in the radio. The continuum of the dust (Figs. 1, 3) is also important in this wavelength domain and can be a problem to get a good signal to noise. For a useful resolution ($\Delta\lambda/\lambda > 10\,000$) large optical (3–10 m class) telescopes on high altitude sites is necessary.

3.4   Electronic spectra of molecules

- Spectral domain: visible to ultraviolet ($\lambda \approx 0.1$–$1\,\mu m$);

- Observable molecules: atoms, ions, small (2–3 molecules) radicals.

In this domain we mostly only see "daughter" molecules which are only made up of a small number of atoms. Parent molecules have electronic transitions generally weaker as they are at lower wavelength were the Solar flux is weaker and the fluorescence mechanism leading to the line emissions is less efficient. This also occurs at wavelengths corresponding to photon energies which are close to the ones necessary to break (photo-dissociate) apart the molecules, so that most large cometary molecules do not show emission lines in the visible.

In the near-UV to visible we see radicals (CN, C_2, OH, CS) and ions while at shorter wavelength mainly only atoms (H, O, C, Ar): at these wavelength the energy of the solar photons absorbed before spontaneous emission ("fluorescence") is generally larger than any molecular binding and polynuclear species cannot survive.

Spectroscopic fine structure of the electronic transition lines usually gets very complex (see example of CO in previous section) because of the numerous sublevels due to rotational and vibrational structure. Very high resolution is again necessary to fully see this structure (e.g. Fig. 11). Modeling the relative intensity of the fine structure lines of those electronic bands (Rousselot et al. 2001) can be very complex, especially when there are forbidden transitions (e.g. C_2 Sects. 3.2–3.3) that spreads radicals on a very large number of energy levels. The total line strength of each electronic transition is however a bit easier to model, but continuum emission from dust must be properly subtracted to determine line intensities, especially in the case of aperture photometry measurements.

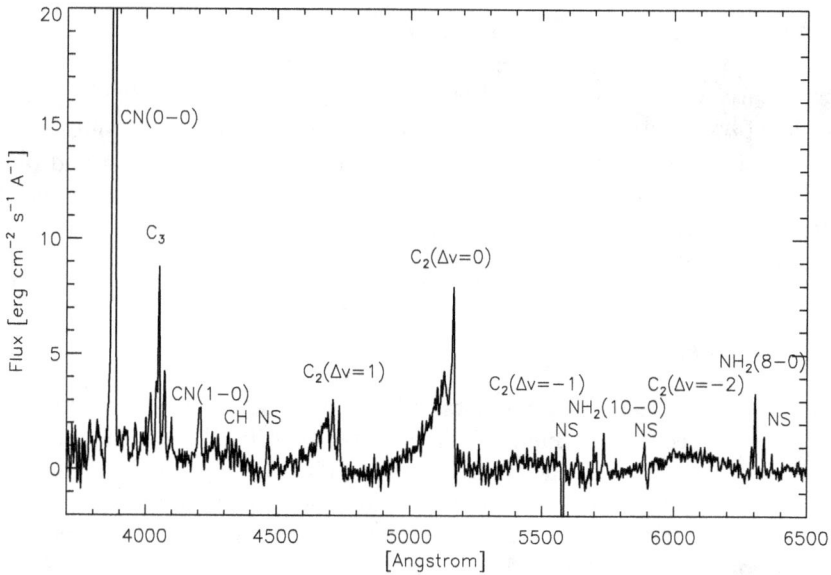

Fig. 10. Visible spectrum of comet C/1995 O1 (Hale-Bopp) obtained with the 1.5 m ESO telescope on the 19th of december 1997. Some night sky (NS) lines have only been partially subtracted (Rauer *et al.* 2003).

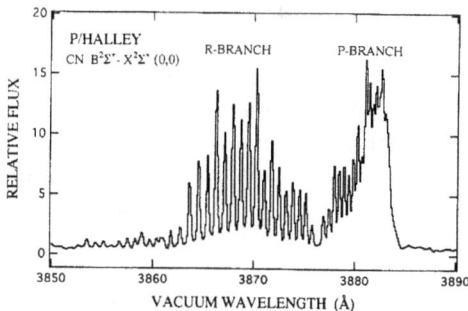

Fig. 11. High resolution spectrum ($R \approx 15\,000$) of the 388 nm CN band in comet 1P/Halley in April 1986 (Kleine *et al.* 1994). Rotational fine structure becomes clear but even a higher resolution $R = \lambda/\delta\lambda = 70\,000$ is necessary to isolate each line (this was use to look for ^{13}CN or C^{15}N lines at $R = 83\,000$ (Arpigny *et al.* 2003)).

3.5 The main cometary lines in the visible

In contrary to the radio and infrared, this wavelength domain is accessible to many observers, especially amateurs. In addition, there are not that many teams of professional astronomers working on cometary spectroscopy, maybe not much more than 100 around the world.

4   Analysis of spectra: Computing physical quantities

In this chapter we will not provide the information on how to reduce observational data (cometary spectra) into calibrated scale of intensity units (typically fluxes in $[\mathrm{Wm}^{-2}\,\text{Å}^{-1}]$ or $[\mathrm{Wm}^{-2}]$). It is any observer duty to reduce his data and remove telluric sky lines; ... We will describe the basis on how to convert these data into physical quantities characterizing the comet activity, outgassing rate, ... as regards to the studied molecules.

4.1   Gas distribution density

This first step is necessary to understand the relationship between the number of molecules observed on the line of sight (Column density N) and the quantity of molecules outgassed by the nucleus every second (Q). To simplify the modeling, we will assume a steady state regime, isotropic coma for the gas and radial expansion at a constant velocity v_{exp}. Variation in those quantities implies additional steps in integration that the interested reader can make.

- Parent molecules are coming directly from the nucleus and destroyed by solar radiation. The photo-dissociation scale-length is Ld ($Ld = v_{exp} \times \tau_d$, τ_d = lifetime, proportional to r_h^2). Then the local density is (Haser model):
$n_{\text{molec.}}(r) = \frac{Q_{\text{molec.}}}{4\pi r^2 v_{exp}} exp(-r/Ld)$.

- Daughter molecules are not coming from the nucleus but from the destruction of a parent molecule with a scale-length Lp, before being themselves photo-dissociated (scale-length Ld). From this simplistic hypothesis we can define another Haser density profile, using the "Haser equivalent" scale-lengths Lp and Ld):
$n_{\text{molec.}}(r) = \frac{Q_{\text{molec.}}}{4\pi r^2 v_{exp}} \frac{Ld}{Lp-Ld}(exp(-\frac{r}{Lp}) - exp(-\frac{r}{Ld}))$.

But, e.g., even if HCN $\rightarrow$ CN by photo-dissociation, $Ld(\text{HCN}) \neq Lp(\text{CN})$, because CN is created with an ejection velocity, isotropically when HCN is broken. So the above formula is not using physically representative scale-lengths, but "equivalent scale-lengths" that make the density profile relatively well representative, provided we use the right parameters. A parameter such as $Lp(\text{CN})$ is including several things such as non pure radial trajectory of CN molecules, photo-dissociation of HCN but also possible contribution of other parent molecules (CH_3CN, HNC, ...) or other sources. Measuring the Lp and Ld parameters directly from the spectral line intensity spatial profiles in various conditions (heliocentric distance, type of comet, ...) is still a useful task.

Data in Table 2, with a supposed heliocentric dependence as $1/r_h^2$ are average values given as indicative, and if the observer can re-measure these values to analyze his data, that is better.

Table 2. Main visible lines of comets.

Molecule	Transition (electronic - vibration band)	wavelength (width)^a	L/N at 1 AU 10^{-20} W	mean relative intensity
C_2	$d^3\Pi_g - a^3\Pi_u \; \Delta v = +1$	473.7 nm (-20 nm)	2.40	0.54
C_2	$d^3\Pi_g - a^3\Pi_u \; \Delta v = 0$	516.5 nm (-30 nm)	4.50	1
C_2	$d^3\Pi_g - a^3\Pi_u \; \Delta v = -1$	563.6 nm (-30 nm)	2.1	0.47
C_2	$d^3\Pi_g - a^3\Pi_u \; \Delta v = -2$	619.1 nm (-30 nm)	0.7	0.15
C_2	$A^1\Pi_u - X^1\Sigma_g^+ \; \Delta v = +1$	1010.0 nm	0.13	0.03
C_2	$A^1\Pi_u - X^1\Sigma_g^+ \; \Delta v = 0$	1210.0 nm	0.05	0.01
C_3	$A^1\Pi_u - X^1\Sigma_g^+ \; v = 000 - 000$	405.2 nm (35 nm)	10.0	0.4
CN	$B^2\Sigma^+ - X^2\Sigma^+ \; v = 1 - 0$	359.0 nm (-4 nm)		
CN	$B^2\Sigma^+ - X^2\Sigma^+ \; \Delta v = 0$	388.3 nm (-4 nm)	2.5–4.5^b	1
CN	$B^2\Sigma^+ - X^2\Sigma^+ \; v = 0 - 1$	421.5 nm (-4 nm)	≈ 0.2	0.07
CN	$A^2\Pi_i - X^2\Sigma^+ \; v = 1 - 0$	914.1 nm ($+15$ nm)	≈ 0.7	0.20
CN	$A^2\Pi_i - X^2\Sigma^+ \; v = 0 - 0$	1093.0 nm ($+15$ nm)	≈ 0.9	0.26
CH	$A^2\Delta - X^2\Pi_r \; v = 0 - 0$	430.5 nm (9 nm)	0.92	
NH	$A^3\Pi_i - X^3\Sigma^- \; \Delta v = 0$	336 nm (10 nm)	0.4–0.9	0.05
OH	$A^2\Sigma^+ - X^2\Pi \; v = 1 - 0$	282.6 nm	0.8–2.7×10^{-3}	0.04
OH	$A^2\Sigma^+ - X^2\Pi \; v = 0 - 0$	306.4 nm ($+5$nm)	15–83 $\times 10^{-3}$	1
OH	$A^2\Sigma^+ - X^2\Pi \; v = 1 - 1$	312.2 nm ($+6$ nm)	1.2–4.2×10^{-3}	0.07
OH	$A^2\Sigma^+ - X^2\Pi \; v = 0 - 1$	346.8 nm	0.05–0.3×10^{-3}	0.01
NH_2	$A^2A_1 - X^2B_1 \; (0,12,0)-(0,0,0)$	515 nm	0.551	
NH_2	$A^2A_1 - X^2B_1 \; (0,11,0)-(0,0,0)$	545 nm	0.279	
NH_2	$A^2A_1 - X^2B_1 \; (0,10,0)-(0,0,0)$	570 nm	0.299	
NH_2	$A^2A_1 - X^2B_1 \; (0, 9,0)-(0,0,0)$	600 nm	0.313	
NH_2	$A^2A_1 - X^2B_1 \; (0, 8,0)-(0,0,0)$	630 nm	0.534	
NH_2	$A^2A_1 - X^2B_1 \; (0, 7,0)-(0,0,0)$	665 nm	0.175	
NH_2	$A^2A_1 - X^2B_1 \; (0, 6,0)-(0,0,0)$	695 nm		
NH_2	$A^2A_1 - X^2B_1 \; (0, 5,0)-(0,0,0)$	735 nm		

Some ion lines, observed far from the nucleus:

Molecule	Transition	wavelength (width)		mean relative intensity
CO+	$A^2\Pi_i - X^2\Sigma^+ \; v = 4 - 0$	379 nm (2 nm)		0.8
CO+	$A^2\Pi_i - X^2\Sigma^+ \; v = 3 - 0$	401 nm (2 nm)		0.9
CO+	$A^2\Pi_i - X^2\Sigma^+ \; v = 2 - 0$	426 nm (2 nm)		1.0
CO+	$A^2\Pi_i - X^2\Sigma^+ \; v = 1 - 0$	455 nm (3 nm)		0.7
CO+	$A^2\Pi_i - X^2\Sigma^+ \; v = 2 - 1$	470 nm (3 nm)		0.6
CO+	$A^2\Pi_i - X^2\Sigma^+ \; v = 1 - 1$	504 nm (4 nm)		0.2
CO+	$A^2\Pi_i - X^2\Sigma^+ \; v = 0 - 1$	549 nm (5 nm)		0.5
CO+	$A^2\Pi_i - X^2\Sigma^+ \; v = 0 - 2$	622 nm (6 nm)		0.5
H_2O+	$A^2A_1 - X^2B_1 \; v = 0,8,0 - 0,0,0$	616 nm		
H_2O+	$A^2A_1 - X^2B_1 \; v = 0,3,0 - 0,0,0$	620 nm		
H_2O+	$A^2A_1 - X^2B_1 \; v = 0,2,0 - 0,0,0$	670 nm		
OH+	$A^3\Pi_i - X^3\Sigma^- \; v = 1 - 0$	336 nm (-4 nm)		
OH+	$A^3\Pi_i - X^3\Sigma^- \; v = 0 - 0$	362 nm (-5 nm)		
OH+	$A^3\Pi_i - X^3\Sigma^- \; v = 0 - 1$	403 nm (-4 nm)		

^{a} A "$-$" (resp. "$+$") sign means that the band extends to shorter (resp. longer) wavelengths from the band head given here;

^{b} $v = 0 - 0$ alone; multiply by 1.08 to take also into account the $v = 1 - 1$ band. (Others are negligible.) – When L/N varies with heliocentric velocity we give the minimum and maximum values it can reach.

The column density will then be (integration on the line of sight at the projected distance ρ from the nucleus):

$$N(\rho) = \frac{2\,Q_{\text{molec.}}}{4\pi v_{exp}}\frac{1}{\rho}\int_{\rho/Lp}^{\infty} K_0(x)dx$$

$$N(\rho) = \frac{2\,Q_{\text{molec.}}}{4\pi v_{exp}}\frac{Ld}{Lp - Ld}\frac{1}{\rho}\int_{\rho/Lp}^{\rho/Ld} K_0(x)dx$$

($K_0(x)$: Modified Bessel function, $\int_0^{\infty} K_0(x)dx = \frac{\Pi}{2}$).

4.2 Molecular "Excitation" (fluorescence process)

This section will provide information on how to convert a line intensity into a number of molecules. The principle is to know the fraction of molecules that emit photons (of energy $h\nu(j)$) at the $\nu(j)$ considered transition frequency. This involves the Einstein coefficient for spontaneous emission of the transition (A_{ij} [s^{-1}]) and the fraction of molecules in the adequate energy state (i) for this emission to happen.

- For a rotational state/transition: the populating of each rotational state will depend on the distance r to the nucleus, as the collision rate will decrease outwards. This is a complex computation that involves two main different process: collisions and solar radiation pumping.

- For vibrational and electronic transitions (for the global intensity of the bands), the dominating process is called fluorescence: a radiative excitation process in which ground energy state molecule absorbs a photon from the Sun radiation field, followed by a spontaneous emission – following the selection rules than can give a different de-excitation route. Thus, the excitation will not depend on the distance from the nucleus (those processes are much faster than collisions), but the fine rotational structure will, as described above. It will be in many cases an "image" of the ground vibrational/electronic state with its rotational structure that is pumped nearly "as is" to higher vibrational/electronic states, as selection rules mostly only allow slight changes ($\Delta J = \pm 1$) in rotational energy levels through radiative pumping.

One major issue in the UV to visible domain is the complexity of the solar spectrum at these wavelength (many absorption features): the velocity of the comet relative to the Sun (r_h), due to the Doppler effect, will change the spectrum seen by the comet. The Sun radiation intensity seen by the comet will then depend on its velocity for the narrow window of each molecular lines. Especially for CN, OH and NH (Schleicher 1983; Schleicher et al. 1987; Meier et al. 1998), the pumping rate will strongly depend on the comet velocity (Swings effect).

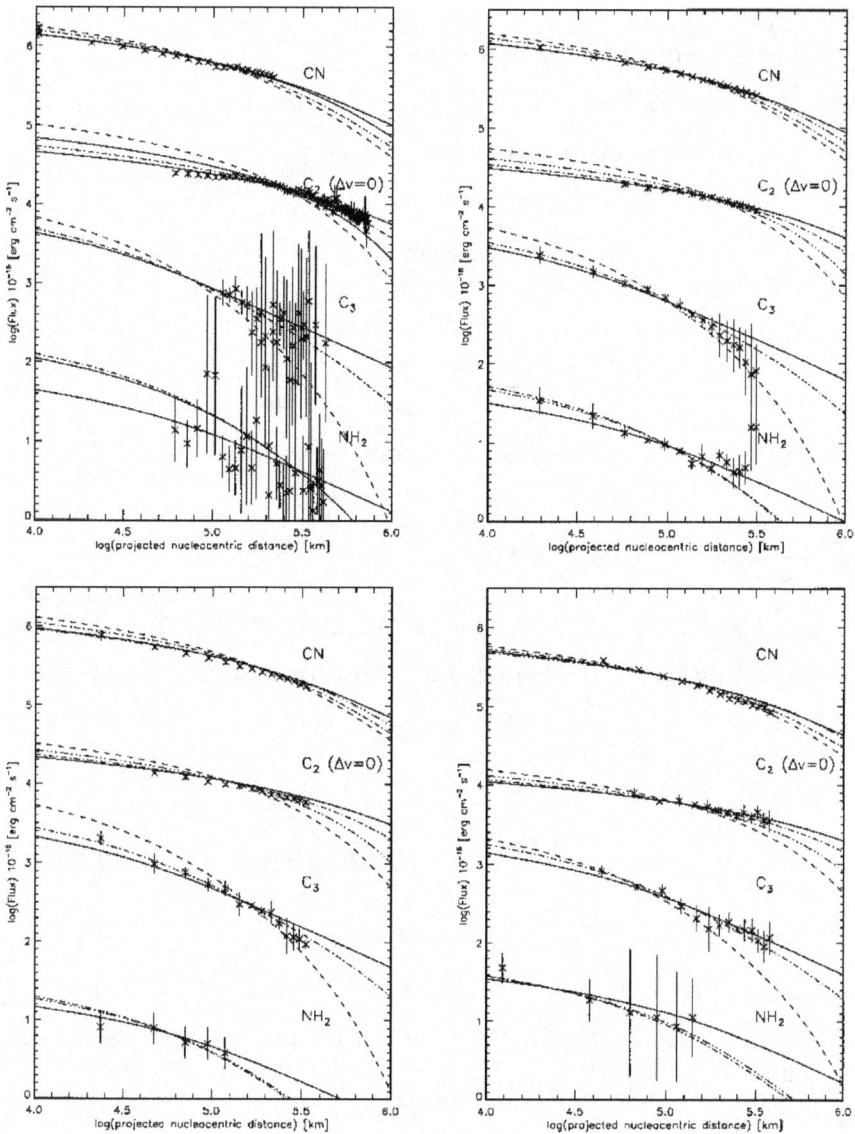

Fig. 12. Radial distribution of lines intensity in the coma of comet C/1995 O1 (Hale-Bopp) for 4 radicals observed (x) in 1997–1998 at ESO. 4 Haser models (lines) have been superimposed (varying Lp, Ld) (Rauer *et al.* 2003).

The variables used to make the conversion between measured flux and column density are:

- F (integrated over the band) is the measured flux [Wm^{-2}];
- $L = 4\pi\Delta^2 F$ total radiated flux in space [Watts = 10^{-7} erg s^{-1}];

- "L/N_{rh}" or $g_r \approx g_0(r_h = 1\,\mathrm{UA}$ (Table 2))$/r_h^2$ is the solar pumping rate of the band (emitted energy in W molecule^{-1});

- S: cometary atmosphere cross section over which the flux is measured [m^2];

- $\Omega = S/\Delta^2$ corresponding solid angle [steradian];

- N: Molecule column density [molecules m^{-2}];

- r_h and Δ are respectively the heliocentric and geocentric distance of the comet (converted into m).

$$N = \frac{L}{SL/N_{rh}} = \frac{4\pi\Delta^2\,Fr_h^2}{Sg_0} = \frac{4\pi Fr_h^2}{\Omega g_0}.$$

4.3 Example

Observation of CN in comet Hale-Bopp on the 19th of December 1997 with the ESO 1.52 m telescope: Figures 10 and 12 (Rauer *et al.* 2003):

- Geometry of the observation: $r_h = 3.78\,\mathrm{AU}$, $\Delta = 3.63\,\mathrm{AU}$, $\dot{r}_h = +18.7\,\mathrm{km\,s^{-1}}$

- Other data: $v_{gaz} \approx 0.65\,\mathrm{km\,s^{-1}}$, $Lp \approx 300\,000\,\mathrm{km}$ (at 3.8 UA), $Ld \approx \infty$

- Pumping rate: $g_0(r_h = 4\,\mathrm{UA}, 1\dot{r}_h = +18.7\,\mathrm{km\,s^{-1}}) = 1.08 \times 3.08 \times 10^{-20}\,\mathrm{W/molecule}$

- $L/N_{rh} = \frac{g_0}{r_h^2} = 0.233 \times 10^{-20}\,\mathrm{W/molecule}$

- Spectrum: Slit of 2.4″, from a line of 0.82″ (1 pixel) at $\rho = 80\,000\,\mathrm{km}$ from center

- $S = 1.34 \times 10^{13}\,\mathrm{m^2}$

- CN line: peak at $23 \times 10^{-17}\,\mathrm{erg\,cm^{-2}\,s^{-1}\,\mathring{A}^{-1}} = 2.3 \times 10^{-19}\,\mathrm{W\,m^{-2}\,\mathring{A}^{-1}}$

- Integrated flux of the line ($\approx 25\,\mathring{A}$): $F = 5.6 \times 10^{-18}\,\mathrm{W\,m^{-2}}$ (Fig. 12b upper right)

- $L = 4\pi\Delta^2\,F = 2.1 \times 10^7\,\mathrm{W}$.

Hence, the column density: $N(\rho = 80\,000\,\mathrm{km}) = 6.7 \times 10^{14}\,\mathrm{m^{-2}}$

$N(\rho) = \frac{Q_{CN}}{2\pi v_{gaz}\,\rho} \int_0^{\rho/Lp} K_0(x)dx \Rightarrow N(80\,000\,\mathrm{km}) \approx 0.7 \times \frac{Q_{CN}}{3.3 \times 10^{11}} = 2.1 \times 10^{-12}Q_{CN}$

So we get Q_{CN} (Hale-Bopp on 19/12/1997)= 3.1×10^{26} molecules per second.

Within 2 AU from the Sun CN/H$_2$O $\approx 0.2\%$ in most cases. So we can extrapolate to a total outgassing rate around 1.7×10^{29} water molecules per second (If not measuring OH) which is about 5 tones per second. In fact, at this distance (3.8 AU) HCN is much more volatile than H$_2$O and we had $Q_{H_2O} = 2.8 \times 10^{28}\,\mathrm{molec.s^{-1}}$ (Crovisier 2000).

5 The observation of dust

Observation of dust may not be as relevant to cometary spectroscopy, but we will just quickly mention it here. Dust signal does depend on wavelength, especially as in the visible one must also avoid confusion with gas spectral features. In the infrared there are event dust (*e.g.* silicates) spectral features to identify, although beyond reach of amateur equipment.

5.1 Distribution of dust

The behavior of dust is somewhat more complicated than molecules, and we will first assume the simpler case where:

- dust emission is uniform and isotropic (no significant jets);
- dust velocity is constant;
- dust grains do not fragment into smaller ones, ...;
- radiation pressure effect is negligible (*i.e.* relatively close to the nucleus).

In this case we can give a density profile following the Haser formalism:

$$n_{\text{dust}}(r) = \frac{Q_{\text{dust}}}{4\pi r^2 v_{dust}} \Rightarrow N(\rho) = \frac{Q_{\text{dust}}}{4v_{dust}} \frac{1}{\rho}.$$

In the reality dust grains are not of unique size and velocity and they can be described with a distribution law: typically the number of grains $n(a)$ of size a increases as $a^{-4} - a^{-3.5}$ to $a^{-4.6}$, considering grains larger than $0.1\,\mu$m. The grain velocity also depends on the sizes: $v_{\text{dust}}(a<1\,\mu\text{m}) \approx v_{\text{gas}} \approx 750\,\text{m s}^{-1}$ $v_{\text{dust}}(a \approx 200\,\mu\text{m}) \approx \frac{v_{\text{gas}}}{10}$.

Most of the dust grains seen in the visible are in the $a = 0.1$ to $10\,\mu$m size range. What will be noticed first on the dust images are the departure from the above $N(\rho) \propto \frac{1}{\rho}$ law, due to radiation pressure and jets effects.

5.2 In the far infrared

As Figures 1 and 2 show, observation of cometary comae in the mid- to far infrared reveals the dust grain thermal emission. Several well calibrated measurements on a wide range of wavelength can be used to measure the black body equivalent temperature of the grains, typically in the 150–400 K range. It can also be used to evaluate the total dust mass present in the cometary coma and dust loss rate of the nucleus Q_{dust}. Since these are performed at a quite different wavelength such measurements are very complementary to visible measurements: they are sensitive to the emissivity and different size domain of the dust particles. Finally dust spectral features can be identified:

- Silicates emission bands at $10\,\mu$m and $20\,\mu$m, whose shapes are sensitive to the ratio of crystalline to amorphous silicates as well as relative abundances of the various forms of silicates (olivine, pyroxene);

- Far from the Sun, water ice absorption bands in the dust grain coma have also been identified (1.5, 2.04 μm) in some active comets.

5.3 In the visible

Cometary dust scatters the solar light. Mie theory can be used to evaluate the scattering efficiency: roughly the larger particles ($a > \lambda$, wavelength) will scatter incoming solar light with an efficiency proportional to $1/\lambda$, while in the case of the smaller particles, it will be proportional to $1/\lambda^4$ following Rayleigh scattering rules.

What can be measured is the reddening (in % variation of the continuum per Å) of the comet dust continuum spectrum in comparison to the incoming solar emission, and its spatial variation. It usually reveals the physical properties of the dust grains. For example, in the case of grain fragmentation as they move away from the nucleus, size distribution will change radially and so can the reddening.

$Af\rho$: This parameter has been invented decades ago by Mike A'Hearn (A'Hearn 1978) to get a parameter easy to measure in order to characterize the dust radiation flux of a comet observed in (any) given photometry aperture. In principle this parameter is a quantity easy (non model dependent) to measure and should be related to the dust production rate. It is measured on a circular aperture (projected radius on the sky ρ) – preferred to rectangular $x \times y$ window – centered on the nucleus. $Af\rho$ = albedo $\times$ filling factor of cometary dust (1 would mean it is fully opaque throughout the field of view selected) $\times$ field radius. Since we saw that $N(\rho) \propto 1/\rho$ then $f \propto \frac{\int N(\rho')\rho' \, d\rho' \, d\theta}{\pi \rho^2}$ is proportional to $1/\rho$. So $Af\rho$ should not depend on the aperture size on which it is measured and be proportional to the total dust production rate Q_{dust}.

In practice, the parameter is evaluated on the calibrated data by: $Af\rho = \frac{\text{Scattered-flux}}{\text{Incoming-solar-flux}} \times \rho$, hence the formulae:

$$Af\rho = \frac{4 \times \pi \Delta^2 \, F}{\pi \times \rho^2 \frac{F_{sun}}{r_h^2}} \times \rho = \frac{(2\Delta \, r_h)^2}{\rho} \frac{F}{F_{sun}} \quad \text{or} \quad Af\rho \approx \frac{(2\Delta \, r_h)^2}{x \times y/(\pi\rho')} \frac{F(\rho')/2}{F_{sun}}.$$

Where Δ, r_h, ρ, ρ', x, y are in $[m]$, F in $[\mathrm{Wm^{-2}}]$, (F_{sun} = solar flux at 1 AU). ρ' is the offset from center and $x \times y$ the rectangular aperture for the approximated formula given by (Cochran 1992).

The $Af\rho$ is the parameter commonly published in scientific publications as the observed quantity, while one would be more interested in the total dust (mass) production rate Q_{dust} to derive the dust-to-gas ratio of the comet. This later one will depend on the modeling, including several parameter as regards to dust properties (size distribution, velocity, albedo, ...) not well characterized.

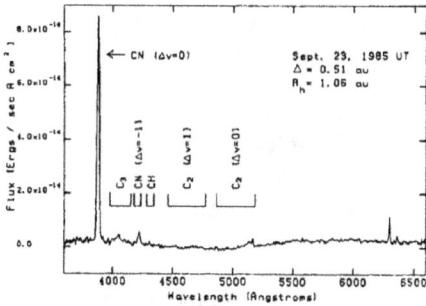

FIG. 1. Spectrum of Comet Giacobini-Zinner obtained 23 September 1985. The most important molecular features are marked. Note the low intensity of the C_2 and C_3 features relative to the CN features.

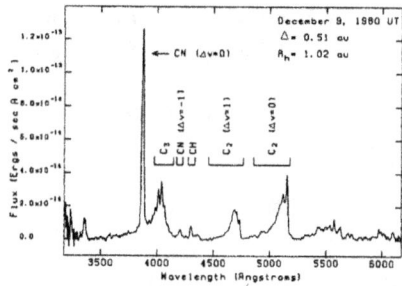

FIG. 2. Spectrum of Comet Tuttle obtained 9 December 1980. This spectrum was obtained at approximately the same heliocentric and geocentric distances as the Giacobini-Zinner spectrum shown in Fig. 1. Note the difference in the C_2 and C_3 molecular emission strengths compared to Fig. 1.

Fig. 13. Two cometary spectra obtained in very similar conditions but showing two comets that have a quite different C_2/CN ratio (Cochran & Barker 1987).

6 What to do with visible cometary spectra?

This section summarizes the main interests of visible cometary spectroscopy.

6.1 Measurable quantities

- When doing spectro-imaging (with a slit): spatial distribution of the molecules and first "qualitative" molecule to molecule or molecule to dust spatial extent comparison;

- Next step is measurement of the molecules (radicals) scale-lengths: "Lp" and "Ld", cf. Section 4.1;

- Comparison between comets of the relative line intensities and lines relative to continuum (e.g. C_2/CN, C_3/CN lines ratios, C_2/dust). In a semi-quantitative way this will tell us quickly about the main characteristics of the comet: C_2 and C_3 depleted or not (cf. example in Figs. 13), large dust to gas ratio or not, ...;

- measurement of the dust reddening (% per Å) throughout the coma.

6.2 Useful to necessary corrections to make reliable comparisons

In a second step, one should do some corrections before making more quantitative comparisons:

- Geometrical effects must be taken into account: Δ the distance to the Earth must be used to convert measured values into physical units [m], and r_h the heliocentric distance is also a parameter on which several parameters such as scale-lengths depend (as r_h^2 or $r_h^{1.5}$ when not evaluated from the observation) – to take into account before quantitative comparisons;

- After taking into account heliocentric distance, one should correct the variation of the pumping rate with **heliocentric velocity** of the comet, in the case of OH and CN, following Tables in Schleicher (1983) and Schleicher & A'Hearn (1988), *e.g.*, in order to compare observations done at different dates or with different comets;

- Finally, from the first step of obtaining a well calibrated spectra (photometrically calibrated data after correction for the various efficiencies of the system, getting rid of night sky lines, using photometric standard stars, ...), converting data into molecular production rates and relative abundances.

If one reaches this last step, then a quantitative comparison between comets is feasible as well as following the evolution of comet with time, distance to the Sun, ...

6.3   With clean (calibrated) data or bigger equipment

Obtaining data of high scientific value is certainly within reach of an experienced observer, well equipped and careful about data acquisition and calibration. At this point data of high scientific value (even of use to the professional community) can be obtained and published:

- Measurements of molecular scale-lengths on the spectro-images: they do not require well photometrically calibrated data, but good quality with good signal-to-noise and some calculation work;

- Precise production rate measurements, as presented in the previous section, but computation with a more sophisticated model of the line pumping process may be useful as a contribution from the professional side;

And finally, here are two aspects of cometary spectroscopy, which really require semi-professional to professional large equipment or experience:

- High resolution ($\lambda/\Delta\lambda > 10\,000$) spectroscopy and analysis of the fine structure of the bands, requiring sophisticated models;

- Very high resolution ($\lambda/\Delta\lambda > 60\,000$) comet spectroscopy that enables to isolate all individual lines of the rotational fine structure: this is required to further measure the ortho/para ratio in NH_2, the $C^{15}N/C^{14}N$, $^{13}CN/^{12}CN$, ...ratios. (Note that the terrestrial ratios are $^{13}C/^{12}C = 1/93$ and $^{15}N/^{14}N = 1/272$). This has been only done with 8-m class telescopes on recent bright comets!

7 Conclusion

The most efficient spectroscopic techniques to study in detail cometary atmospheres are beyond amateurs means: radio and infrared requires large expensive equipment and even in the visible sensitive programs also require large equipment. But, on the other hand, the comet investigations in these domain are quite limited: the teams of scientists/observers are very small (on the order of 100 or so all over the world for all spectroscopic studies of comets) and corresponding observing time with the large facilities are limited.

Comets are variable targets and differ from each other, so that wide coverage in targets and time with less deep studies is also essential. Amateurs can play an essential role here: limited equipment (20 cm class telescope with mid-to-low resolution optical spectrometer $R \approx 200\text{--}1000$) can be enough for a useful work. The experienced amateur careful about data quality and calibration can provide very valuable data, especially as very few professional would do the same. The radicals (C_2, C_3, CN, NH_2 up to OH in the near UV) that can be detected and are also very useful to monitor comet activity. We have seen that it is possible to retrieve quantitative information and the experienced amateur can also provide valuable information in a short time that professional community would appreciate to plan extensive investigations. In the future we can even expect professional–amateur collaborations to publish amateur work in scientific refereed publications.

References

A'Hearn, M.F., 1978, AJ, 219, 768

A'Hearn, M.F., 1982, in COMETS, ed. Laurel L. Wilkening (University of Arizona press), 433–460

A'Hearn, M.F., Millis, R.L., Schleicher, D.G., Osip D.J., & Birch, P.V., 1995, Icarus, 118, 223

Arpigny, C., Jehin, E., Manfroid, J., et al., 2003, Science, 301, 1522

Biver, N., 1996–2002, in "Interaction Rayonnement-Matière dans les atmosphères Planétaires et Cométaires", ed. G. Moreels (Besançon Observatory), 205

Bockelée-Morvan, D., Biver, N., Colom, P., et al., 2004, Icarus, 169, 113

Cochran, A.L., & Barker, E.S., 1987, AJ, 92, 239

Cochran, A.L., Barker, E.S., Ramseyer, T.F., & Storrs, A.D., 1992, Icarus, 98, 151

Crovisier, J., & Encrenaz, T., 1995, "Les Comètes, Témoins de la Naissance du Système Solaire" (Belin-CNRS Editions)

Crovisier, J., 1996–2002, in "Interaction Rayonnement-Matière dans les atmosphères Planétaires et Cométaires", ed. G. Moreels (Besançon Observatory), 137

Crovisier, J., 2000, in IAU Symp. 197 "Astrochemistry: From molecular Clouds to Planetary Systems", ed. Y.C. Mihn & E.F. van Dishoeck (ASP), 461

Festou, M.C., & Zucconi J.-M., 1984, A&A, 134, L4

Festou, M.C., Rickman, H., & West R.M., 1993, A&AR, 4, 363,

Festou, M.C., Rickman, H., & West R.M., 1993, A&AR, 5, 37

Kleine, M., Wyckoff, S., Wehinger, P.A., & Peterson, B.A., 1994, AJ, 436, 885

Laffont, C., "Etude d'émissions gazeuses dans les régions internes de 3 comètes: Halley, C/1996 B2 (Hyakutake) et C/1995 O1 (Hale-Bopp)" Université Paris 6, Ph.D. Thesis, 1998

Lecacheux, A., Biver, N., Crovisier, J., *et al.*, 2003, A&A, 402, L55

Meier, R., Wellnitz, D., Kim, S.J., & A'Hearn, M.F., 1998, Icarus, 136, 268

Mumma, M.J., McLean, I.S., DiSanti, M.A., *et al.*, 2001, AJ, 546, 1183

Rauer, H., Helbert, J., Arpigny, C., *et al.*, 2003, A&A, 397, 1109

Rousselot, P., Arpigny, C., Rauer, H., *et al.*, 2001, A&A, 368, 689

Schleicher, D.G., 1983, "The fluorescence of cometary OH and CN" University of Maryland, Ph.D. Thesis

Schleicher, D.G., Millis, R.L., & Birch, P.V., 1987, A&A, 187, 531

Schleicher, D.G., & A'Hearn, M.F., 1988, AJ, 331, 1058

Schleicher, D.G., Millis, R.L., & Osip D.J., 1991, Icarus, 94, 511

Tegler, S.C., Campins, H., Larson, S., *et al.*, 1992, AJ, 396, 711

Weaver, H.A., & Feldman, P.D., 1992, in " Science with the Hubble Space Telescope", ed. P. Benvenuti, & E.J. Schreier, ESO Conf. Workshop Proc., 44, 475

Proceedings of the first international conference on comet Hale-Bopp: Earth, Moon and Planets, 78 (Kluwer Academic Publishers, 1997–1999)

Proceedings of the first international conference on comet Hale-Bopp: Earth, Moon and Planets, 79 (Kluwer Academic Publishers, 1997–1999)

Proceedings of the international conference "Cometary Science after Hale-Bopp": Earth, Moon and Planets, 89 (Kluwer Academic Publishers, 2002)

Proceedings of the international conference "Cometary Science after Hale-Bopp": Earth, Moon and Planets, 90 (Kluwer Academic Publishers, 2002)

Astronomical Spectrography for Amateurs
J.-P. Rozelot and C. Neiner (eds)
EAS Publications Series, **47** *(2011) 189–214*

SPECTROMETRY OF NEBULAE

A. Acker[1]

Abstract. Nebular emission lines are easy to observe, and their spectrum contains a lot of information. We explain the mechanisms of production of the emissions, and the relation between the intensity of the recombination and forbidden lines, and the physical parameters of the objects. A gallery of emission lines spectra is presented, and a rough analysis will clarify their differences. The case of Planetary Nebulae will be developed, in order to determine the extinction constant, the plasma parameters (electron density and temperature), the chemical abundances, and also the properties of the central star (temperature, mass, stellar wind velocity, age).

1 Introduction

Nebulae are fascinating astronomical objects, varied, not easily classifiable, and interesting in all spectral regions. Emission lines are emitted by all nebulae of low density heated and ionised by hot stars.

Emissive circumstellar shells envelop stars at different stages of their history:
- very young stars – such as TTauri stars – embedded in their proto-stellar nebula;
- Wolf-Rayet (WR) stars surrounded by a shell produced by large mass-loss in the final stages of their evolution;
- very old stars having ejected gas and dust at the end of their life – such as planetary nebulae (PN) around low mass stars, "Ring nebulae" around massive WR stars, and supernovae remnant (SNR) around neutrons stars (degenerated very massive stars).

Extended nebular regions are found in the interstellar medium. Ionized hydrogen regions (H^+ or HII Regions) are generally mixed with dusty clouds, and are the theater of active starbirth.

Some galaxies show bright emission lines in their spectrum: galaxies with starburst regions, or with a large number of massive WR stars (WR galaxies), or galaxies with active mass loss in their bulge (AGN).

---

[1] Observatoire de Strasbourg, 11 rue de l'Université, 67000 Strasbourg, France;
e-mail: `agnes.acker@astro.unistra.fr`

© EAS, EDP Sciences 2011
DOI: 10.1051/eas/1147007

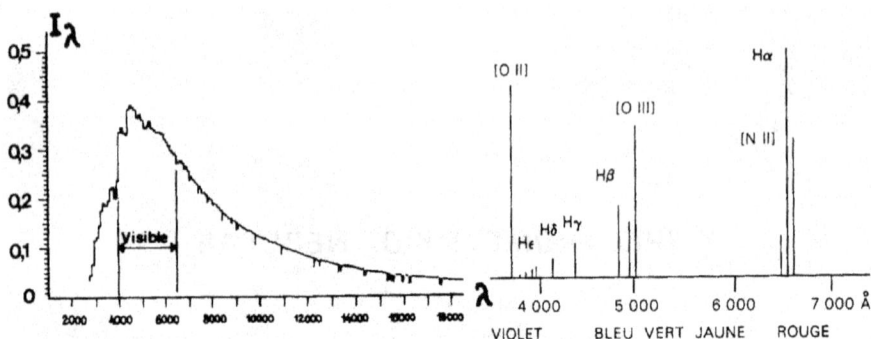

Fig. 1. Spectrum of the Sun –left– and of a nebula –right– (from Acker 2005).

2   Nebular emission lines – Recombination and forbidden lines

With the introduction, some years ago, of new types of high sensitivity detectors (such as CCD cameras), it became possible to record spectra of very faint objects, which could not be investigated during the photographic era. Therefore, the spectrometric techniques are now open to an increasing number of observers, including amateurs. Emission lines objects are the best candidates for productive work, because the whole energy they radiate is concentrated in very narrow spectral domains: the emission lines.

2.1   The emission lines

In 1830, Fraunhofer observed the spectrum of the Sun (see also the chapter of J.P. Rozelot in this volume), and discovered fine black lines on the bright background (26 000 observed in the visible region). A few years latter, in 1860, a German physicist, Gustav Kirchhoff published the laws describing the production of three kind of spectra:
- a continuous spectrum is emitted by a hot, very dense gas or a hot solid incandescent object;
- discrete bright radiations are emitted by a hot, diffuse gas (emission lines);
- discrete radiations are absorbed (absorption lines) in the continuous spectrum by a cool, diffuse gaz in front of the continuous spectrum emitter.
The important conclusions of these observations concern the physical properties of the sources: (i) a continuous spectrum depends on the temperature of the source, which appears bluer (shorter wavelength) if the temperature of the source increases; (ii) every chemical element produces its specific spectral lines feature (emission or absorption), and could therefore be identified by its spectral line fingerprint.

The spectra of astronomical objects appear like the Sun's spectrum (typical star spectrum, where absorption lines appear over the continuum), or like a nebular spectrum, as shown on Figure 1.

The Hydrogen spectrum. The description of the wave nature of light was completed in 1920 by Einstein in the framework of the quantum mechanics (Einstein obtained the Nobel Price for his explanation of the photoelectric effect): electromagnetic light waves are carrying energy in a stream of massless particles, called photons in 1926 by G. Lewis. Each photon of wavelength λ transports the Planck's quantum energy:

$$E_{photon}(J) = \frac{hc}{\lambda}.$$

This energy is equal to 4×10^{-19} J for a yellow photon, with $\lambda = 0.5 \times 10^{-6}$ m, the Planck constant $h = 6.626 \times 10^{-34}$ $J{\cdot}s$ and c the light velocity.

The particle-wave duality of light was applied to the structure of the atom – as proposed by Niels Bohr in 1915 – in order to explain the production of discrete spectral lines. The angular momentum of the electron orbiting around the nucleus was quantized, which means that the angular momentum of the hydrogen atom must be equal to integral multiples of Planck's constant divided by 2π.

By using the Coulomb's and Newton's laws, Bohr demonstrates that in the hydrogen atom the electron orbits at specific distances from the proton, all orbits having allowed energies (depending on the Planck's constant, the charge of the electron, and the masses of the proton and the electron):

$$E_n(eV) = \frac{-13.6}{n^2}(\text{where } n \text{ is the principal quantum number}).$$

In the lowest orbit (the ground state) the energy of the electron is -13.6 eV (this is the value of the ionization potential). In the second orbit (first excited state), the energy is -3.4 eV. When an electron makes a transition from a high (h) to a low (l) orbit, a photon is emitted, that carries away an energy equal to the difference of energy $E_h - E_l$:

$$E_{photon} = \frac{hc}{\lambda} = 13.6 \left[\frac{1}{n_l{}^2} - \frac{1}{n_h{}^2} \right]$$

$$\frac{1}{\lambda} = \frac{13.6}{hc} \left[\frac{1}{n_l{}^2} - \frac{1}{n_h{}^2} \right] = 109677.5 \left[\frac{1}{n_l{}^2} - \frac{1}{n_h{}^2} \right] \text{cm}^{-1}.$$

This expression gives the generalized Balmer formula for the hydrogen lines and confirms the empirical relation found by Johan Balmer in 1885: $n_l = 1$ for the Lyman series (in the UV), $= 2$ for the Balmer series (in the visible), $= 3$ for the Paschen series (in the IR) (Fig. 2).

The recombination lines. The recombination of an electron with a H^+ ion leads to emission lines.
The various energy levels will be calculated in two cases:
- Case A: "low" density, *e.g.* the nebula is optically thin in all lines
- Case B: "high" density, *e.g.* the nebula is optically thick in the Lyman lines, a more realistic situation that case A.

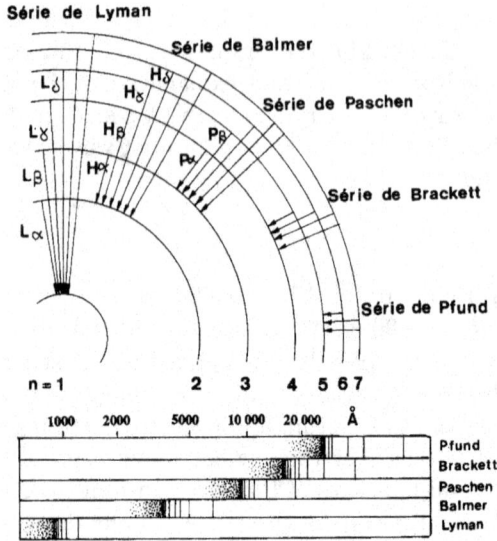

Fig. 2. Hydrogen emission lines series (from Acker 1992).

The lifetime on an excited level is very short ($\approx 10^{-8}$ sec). The emissivity j could be expressed as follows:

$$j = \frac{h\nu}{4\pi} N_e N_H . \alpha(N_e, T_e)$$

α: coefficient of effective recombination $\propto 1/T_e$.

N_e and T_e: electron density and temperature.

N_H: atomic hydrogen density.

Very high energy is needed to ionize and excite hydrogen and helium. But the abundance of these elements is very high, therefore the visible nebular spectrum shows bright hydrogen Balmer lines, dominated by the $H\alpha$ line (see Fig. 1).

All nebular spectra show recombination lines (the brightest are seen in the red spectral range) and *forbidden* lines marked by the element inside bracket (dominated in the visible domain by a green doublet).

The forbidden lines (collisional excitation). In 1864-65, William Huggins examined (for the first time) the spectrum of a nebula, NGC 6543, and was unable to identify the green doublet – never observed in terrestrial laboratories, and unknown from the theoretical spectroscopists. Therefore the green doublet was ascribed to a new element called "nebulium" (following the story of "helium" discovered in 1859 in the Sun, before its observation in a laboratory).

In 1928, Bowen identified the "nebulium" lines as due to transitions between the low energy levels in the ground configuration of O++, transitions excited by electrons collisions. But the radiative transition probability is very small. In terrestrial laboratories (if the gas density is too high), the excited level is mostly

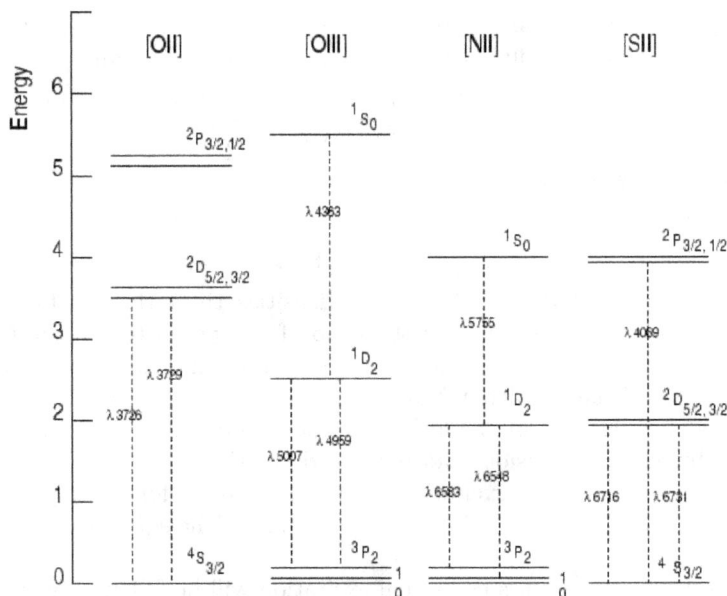

Fig. 3. OII, OIII, NII, SII Energy levels (eV).

depopulated by electron collisions before production of the radiative transition, and therefore the lines were called *forbidden*.

The total rate of collisional desexcitations per unit volume (cm^3 s^{-1}) is related to the quantum of mechanical collision strength between 2 levels, Ω (having a value of about 1):

$$N_e n_2 q_{21} = 8.6 \times 10^{-6}.N_e n_2.\frac{N_e n_2}{T e^{1/2}}\frac{\Omega_{12}}{g_2}. \tag{1}$$

The rate of excitations is given by:

$$N_e n_1 q_{12} = N_e n_1 \frac{g_2}{g_1} q_{21}.e^{-\chi_{12}/kT}. \tag{2}$$

χ_{12}: difference of energy between the two levels
k: Boltzmann constant
n_i: population of the level i
q_{21} and q_{12}: collisional desexcitation probability and collisional excitation probability of the levels 1 to 2
g_1: statistical weight of the level 1
N_e and T_e: electron density and temperature.

For each 2 levels ion a collisional excitation will be followed by a radiative desexcitation, with a corresponding line emissivity: $j_{21} = n_2 A_{21} h\nu_{21} = N_e n_1 q_{12} h\nu_{21}$.

The Figure 3 shows the lowest levels of OII, OIII, NII and SII.

In a high density gas, many collisional desexcitations occur (q_{21} not equal to zero), with the following equilibrium: $n_2(N_e q_{21} + A_{21}) = N_e n_1 q_{12}$

$$\frac{n_2}{n_1} = \frac{N_e q_{12}}{A_{21}} \left[\frac{1}{1 + N_e q_{21}/A_{21}} \right]. \tag{3}$$

The line emissivity becomes:

$$j_{21} = N_e n_1 q_{12} \frac{h\nu_{21}}{4\pi} \left[\frac{1}{1 + N_e q_{21}/A_{21}} \right].$$

For $N_e \rightarrow \infty$ we find $A_{21} \rightarrow 0$. At high densities the forbidden lines are not produced, as they are quenched by the ratio of the permitted to the forbidden transition rates, a ratio that is at least 6 orders of magnitude smaller than 1 (this explain the *forbidden* denomination of these lines). The levels that would be populated at low density for the forbidden lines are depopulated at high density by the combination of collisions and permitted transitions.

In a low density gas, the excitation leads always to the forbidden line emission through radiative desexcitation (q_{21} being negligible). The equilibrium is given by: $n_2 A_{21} = N_e n_1 q_{12}$.

For every two levels ion, a collisional excitation will be followed by a radiative desexcitation, with the following line emissivity:

$$j_{21} = n_2 A_{21} \frac{h\nu_{21}}{4\pi} = N_e n_1 q_{12} \frac{h\nu_{21}}{4\pi}.$$

For $N_e \rightarrow 0$, the emissivity of the forbidden lines appears by many orders of magnitude higher than those of recombination lines for two reasons: (1) As the nebular density is low, the collisionally desexcitation is very limited. (2) The low levels of the ions are separated by very low energy quantum (a few kT_e) leading to emission lines appearing in the visible or near ultraviolet ranges, whereas the *permitted* transitions imply higher levels, and therefore are not easy to excite, being produced in the far UV range (see Fig. 4).

Forbidden lines are the best cooling vectors of the gas. As shown on Figure 4, the emissivity of bright nebular lines reaches a maximum for nebular temperatures higher than 10 000 K. The CIV line appears in the UV range, whereas the [SIII] line is emitted in the IR range. The [OIII] green doublet is the prominent emission in the visible range.

The lifetime on metastable levels is very long (seconds to hours) leading to a very small probability of transition: $P_1(s^{-1}) = 0.03$. The probability that electron leaves the metastable level due to a collision essentially depends on the density: $P_2(s^{-1}) = 3.10^{-4} \times \frac{N_e}{\sqrt{T_e}}$.

Forbidden lines are emitted until $P_1 = P_2$, which means $\frac{N_e}{\sqrt{T_e}} = 100$.

This ratio is realized for the usual conditions inside planetary nebulae or high excited HII Regions: density $\simeq$ temperature $\simeq$ 10 000.

If the temperature decreases, the forbidden lines are no longer emitted. As their very intense lines provided high cooling rate, the temperature will then increase. When $\frac{N_e}{\sqrt{T_e}} \leq 100$, the lines will be re-emitted, etc.

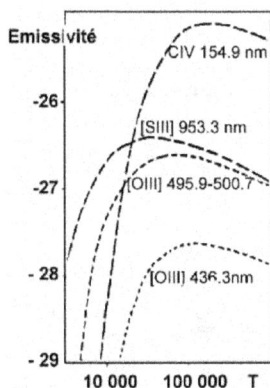

Fig. 4. Emissivity of nebular lines (in W/cm^3, per ion and per incident electron, on a log scale), *versus* electron temperature for a weak density (some $10^2/cm^3$) – (from Grazyna Stasinska, private communication).

2.2 Determination of the plasma parameters by using forbidden lines

The nebular temperature using the [OIII] lines. The [OIII] ion is characterized by excitation potentials of 2.5 eV and 5.3 eV between the low levels ^3P and the levels ^1D and ^1S (Fig. 3). The transition from $^1S_0$ to $^1D_2$ gives the forbidden line at 4363 Å. Transitions from $^1D_2$ to $^3P_{1,2}$ give the 5007 Å and 4959 Å forbidden lines – having a constant intensity ratio of 2.9. If the nebular temperature increases, the ratio of the number of ions in the $^1S_0$ state to the number of ions in the $^1D_2$ state increases, and as a consequence the 4363 Å line becomes brighter compared to the increasing of the 5007 and 4959 Å lines. Therefore, it is possible to estimate the electron temperature of the nebula by measuring the ratio R:

$$R = \frac{j(\lambda 5007) + j(\lambda 4959)}{j(\lambda 4363)} = \frac{j(^1D_2 \to^3 P_2) + j(^1D_2 \to^3 P_1)}{j(^1S_0 \to^1 D_2)}.$$

After several calculations, using numerical value of every symbol, we obtain:

$$R = 8.32 \times e^{(3.29 \times 10^4)/T_e} \times \frac{1}{K}$$

with $K = 1 + 4.5 \times 10^{-4} \times \frac{N_e}{\sqrt{T_e}}$.

R appears highly dependent from the temperature. For usual nebular densities, $K \simeq 1$, and the relation can be approximated to give

$$T_e = \frac{3.29 \times 10^4}{\ln(R/8.3)} \quad \text{(Fig. 5)}.$$

For cooler nebular regions, the temperature could be done by the ratio of the [NII] forbidden lines intensities: [NII] $\frac{j(\lambda 6583) + j(\lambda 6548)}{j(\lambda 5755)}$. Note that the ratio of the red lines $\frac{j(\lambda 6583)}{j(\lambda 6548)}$ has a constant value of about 3.

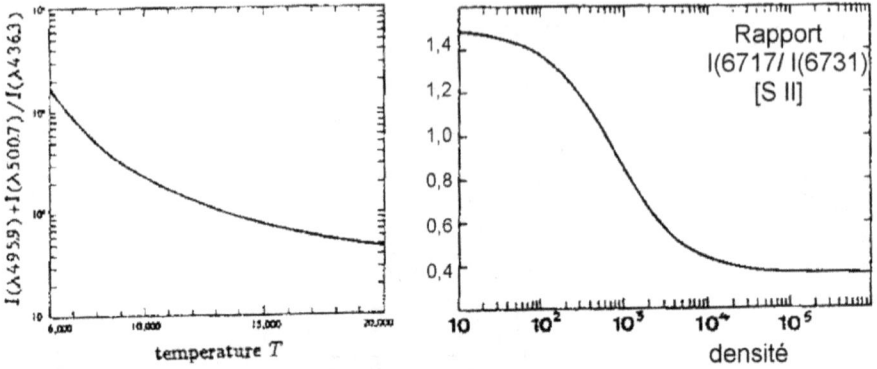

Fig. 5. *Left*: intensity ratio of $[OIII]$ lines as a function of electronic temperature (K) for low density nebulae (from Osterbrock 1974). *Right*: intensity ratio of $[SII]$ lines as a function of electronic density assuming $T = 10^4$ K (from Osterbrock 1974).

The nebular density using the [SII] lines. The ratio $\frac{j(\lambda 6716)}{j(\lambda 6731)}$ of the red [SII] doublet will allow the determination of the density, as both lines nearly have the same excitation and therefore their ratio is temperature independent. The excitation equilibrium includes collisional desexcitation.

If the highest energy level is $2a$ and the lowest $2b$, we obtain:

$$\frac{j(\lambda 6716)}{j(\lambda 6731)} = \frac{n_{2a} A_{2a1} h\nu_{2a1}}{n_{2b} A_{2b1} h\nu_{2b1}} \approx \frac{n_{2a} A_{2a1}}{n_{2b} A_{2b1}}. \tag{4}$$

Equation (3) gives: $\frac{n_{2a}}{n_1} = \frac{N_e q_{12a}}{A_{2a1}}\left[\frac{1}{(1 + \frac{N_e q_{2a1}}{A_{2a1}})}\right]$, for index a, and the same expression stands for index b.

By combining these equations, we obtain a relation between the line ratio and the density:

$$\frac{j(\lambda 6716)}{j(\lambda 6731)} = \frac{q_{12a}}{q_{12b}} \cdot \left[\frac{(1 + \frac{N_e q_{2b1}}{A_{2b1}})}{(1 + \frac{N_e q_{2a1}}{A_{2a1}})}\right].$$

All other levels are not taken into account as the usual nebular temperatures are too low and cannot excite higher levels.

- For very low density, $N_e \ll \frac{A_{2a1}}{q_{2a1}}$:

$$\frac{j(\lambda 6716)}{j(\lambda 6731)} = \frac{q_{12a}}{q_{12b}} = \frac{\Omega_{2a1}}{\Omega_{2b1}} \cdot e^{\chi_{2a2b}/kT} \approx \frac{\Omega_{2a1}}{\Omega_{2b1}} = \frac{g_{2a}}{g_{2b}}.$$

For the levels $^2D_{5/2}(= 2a)$ and $^2D_{3/2}(= 2b)$, $g_{2a} = 6$ and $g_{2b} = 4$ leading to a ratio $= 3/2$ at the low density limit.

- For the highest densities, $N_e \gg \frac{A_{2a1}}{q_{2a1}}$:

$$\frac{j(\lambda 6716)}{j(\lambda 6731)} = \frac{g_{2a} A_{2a1}}{g_{2b} A_{2b1}} \approx 0.38 \quad \text{(Fig. 5)}.$$

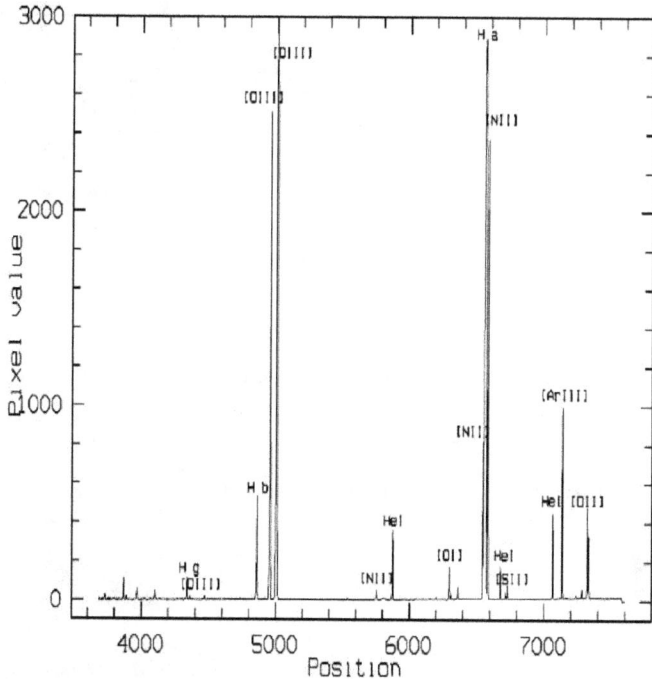

Fig. 6. Spectrum of a typical planetary nebulae, PN G055.5-00.5 (as observed by the author on the 152-cm European Southern Observatory telescope).

The most important "plasma diagnostic" lines. The most important lines in the visible spectral range for the diagnostic of the plasma are the following (see Fig. 6):

1. The *Interstellar extinction* is obtained thanks to the Balmer decrement, essentially through the ratio of the Hα to the Hβ lines (observed ratio compared to the theoretical one, 2.85). By using this extinction constant, all lines must be de-reddened in order to determine the physical and chemical parameters of the nebula.

2. The *Electron Temperature* is obtained by the ratio R of the green [OIII] doublet to the λ4363 line.

    For cooler nebular regions, the temperature should be estimated through the ratio of [NII] lines intensities $\frac{j(\lambda 6583)+j(\lambda 6548)}{j(\lambda 5755)}$.

3. The *Electron Density* is obtained by the ratio of the [SII] doublet $\frac{j(\lambda 6716)}{j(\lambda 6731)}$. For spectra with a well observed violet range (and a fairly high resolution), one can also use the [OII] doublet $\frac{j(\lambda 3729)}{j(\lambda 3726)}$.

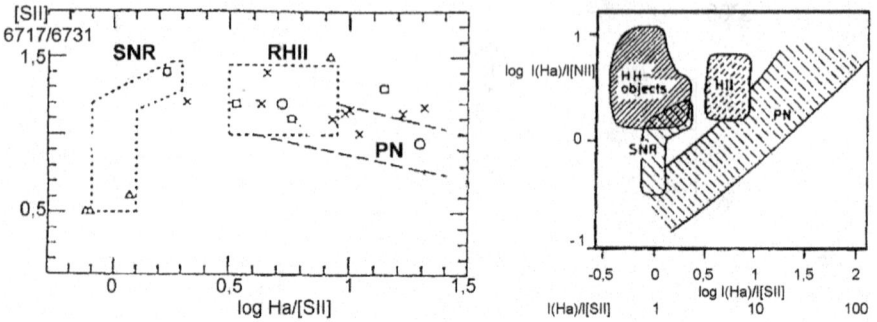

Fig. 7. *Left*: relation between the intensity ratio of the [SII] lines, and the Hα and [SII] lines for various nebular objects. *Right*: relation between the intensity ratio of the Hα and the [SII] and [NII] lines for various nebular objects (from Sabbadin *et al.* 1984, 1986).

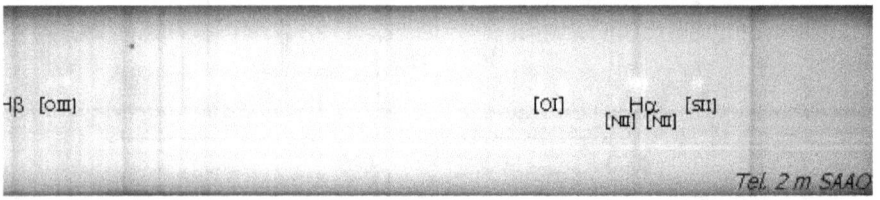

Fig. 8. 2D-spectrum of a faint planetary nebula, performed by Quentin Parker at the 2 m. SAAO telescope, during a full Moon night. – The main nebular lines are identified.

4. The *Excitation-class* of the nebula is estimated by the Helium II λ4686 line, appearing as strong as the Hβ line for high excited nebulae with a very hot stellar nucleus (see Fig. 19). Neutral regions are marked by the presence of intense [OI] lines especially the λ6300 line.

Note that the [SII] doublet is especially interesting, because its global intensity value (relatively to the Hα line) is discriminating for the separation of photoionised nebulae and very dense nebulae dominated by shocks (SNR or Herbig-Haro objects or AGN galaxies) (Fig. 7).

3   A gallery of emission lines objects

Emission lines objects are in most cases faint nebular sources, such as planetary nebulae or galaxies. Therefore, the emission coming from these objects could be difficult to extract with small telescopes. Figure 8 shows the CCD-2D spectrum of a faint, very small, planetary nebula belonging to the galactic bulge. Permitted and [forbidden] nebular lines are identified. In order to extract the nebular spectrum from this 2D image, we have to subtract (1) the cosmic rays (appearing as bright dots over a few pixels), (2) the solar spectrum diffused by the moon (bright continuum and dark absorption lines covering the whole

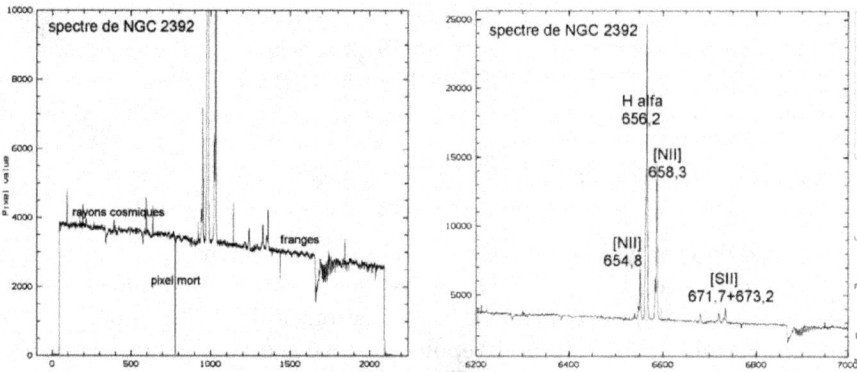

Fig. 9. *Left*: observed spectrum of the planetary nebula NGC 2392 (note the cosmic rays lines, the death pixel, the CCD interference fringes on the red part of the spectrum). *Right*: reduced spectrum of NGC 2392 (OHP-tel. 152 cm, AURELIE spectrograph, R = 5000, exposure 600 sec – Observation DEA Strasbourg).

field), (3) the bright night sky emission lines covering the whole field. (Note that these lines could be identified thanks an applet created by Dr Joachim Köppen: http://astro.u-strasbg.fr/ koppen/applet). (4) In the reduction procedure we had to take into account the presence of field stars (bright continuum), and (5) the CCD defaults (see the chapters by C. Buil and V. Desnoux).

Reduction of the spectra. Spectrometric observations were performed at the 152 cm telescope of the *Observatoire de Haute-Provence (OHP)* equipped with the spectrograph *Aurelie*, in the framework of a 5 nights run performed by the author and her Strasbourg University pre-doctoral students. The spectrograph offers a spectral range of 900 Å with a spectral resolution power of about 5000 ($\frac{\lambda}{\delta\lambda}$). We selected the red spectral range, less affected by interstellar extinction, and leading immediately to the estimation of the nebular density. The obtained 1D spectra of a series of emission lines objects are shown on Figure 10.

Each spectrum was reduced by using MIDAS, a software created by the *European Southern Observatory* (*Munich Image Data Analysis Software* available on a CD-ROM proposed on the ESO web site), in order to transform the observed spectrum into a reduced spectrum (example of the planetary nebula NGC 2392, Fig. 9).

Calibration spectra were performed each night and used as follows:

- The *OFFSET* spectrum will be substracted from each spectrum. In order to avoid to introduce noise, a median spectrum on a series of 25 spectra was used.

- The *Flat Field* spectrum will allow to eliminate all CCD defaults. It is the continuous spectrum of a tungsten lamp, divided by his best smoothed fit, in order to obtain the *flat field spectrum normalized to 1, FFN*. This spectrum shows the difference to 1 due to (i) the irregular response of each pixel, (ii) the interference fringes produced by the CCD. Each astronomical spectrum (corrected from the offset and the cosmic rays) will be divided by *FFN*.

– The *Wavelength calibration* spectrum of Argon and Thorium lamps allows the transformation of the pixel scale into the wavelength scale.

The obtained spectra will then be flux calibrated, by applying corrections concerning (i) the extinction and reddening caused by terrestrial atmosphere (by using specific laws and Tables adapted to the observation site); (ii) the global response of the instrument. In this perspective, one to three spectrophotometric standard stars are observed during each night. The reduced spectrum of each standard star is divided by the flux calibrated spectrum of this star available in published Tables, in order to obtain the *response curve* of the current part of night.

In brief, the reduced spectrum is obtained following 6 steps:
- suppress cosmic rays and night sky lines;
- substract the *OFFSET*;
- divide by *FFN*;
- calibrate in waveleghts;
- correct from earth atmosphere effects;
- divide by the *response curve*.

Emission-lines objects.   Spectra are shown on Figure 10.

Spectra of very young stars
The stars **T Tau and CQ Tau** represent a stage preceding the equilibrium on the main sequence. The stars show a stellar continuum with absorption lines, in particular Hα, and remains embedded in its proto-stellar nebula showing all typical emission lines (Hα, the [SII] doublet, [NII], [OI], see Fig. 10).

Spectra of stars at the end of their evolution
The **WR stars** are hot stars in the final stage of their life. WR 137 is a massive hydrogen-poor star, exposing the helium burning products carbon and nitrogen. The star lost its residual hydrogen-rich envelope at the stage *Luminous Blue Variable* (LBV). The stellar lines are enlarged by violent mass ejection: the FWHM of 59 Å for the two bright lines (see Fig. 10) corresponds to a stellar wind of about $3000 \, km \, s^{-1}$. The LBV nebula in expansion remains visible through the Hα line seen on the blue top of the large stellar WR line. The nebular line is split into two narrow components, revealing an expansion velocity of $175 \pm 25 \, km \, s^{-1}$.

Planetary nebulae (PN) are ejected by low mass stars after the *Asymptotic Giant Branch* (AGB) stage (see next section). The spectrum is dominated by intense nebular lines, over a faint blue stellar continuum (see the spectra of NGC 6543 and NGC 2392, Fig. 10 and Fig. 9). Figure 10 shows a longer exposed nebular

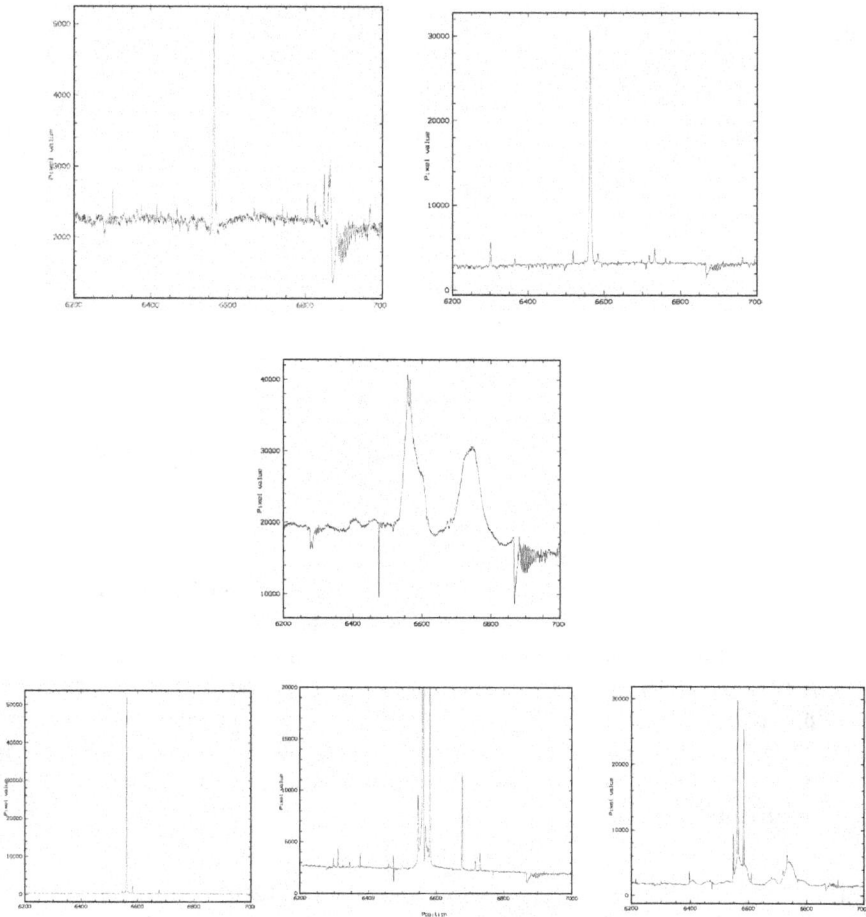

Fig. 10. (First part). The red part of the spectrum of (1a) the proto-star CQ Tau, exposure 600 sec – (1b) the proto-star T Tau, exposure 600 sec – (2) the star WR 137 – (3) the planetary nebula NGC 6543 (a) exposure 60 sec – (b) exposure 10 min, with a saturated Hα line – (3c) the planetary nebula NGC40, exposure 3 min – All spectra were calibrated in relative fluxes. On all spectra, note the strong absorption feature at 5870 Å due to the water molecule of the terrestrial atmosphere. The spectra were performed in March 2003 by pre-doctoral students from Strasbourg Observatory, at OHP-tel. 152 cm, AURELIE spectrograph, R = 5000.

spectrum of NGC 6543, with a saturated Hα line (which cannot be used for the analysis). The [SII] lines ratio indicate a density of a few 10^3 cm^{-3}.

Some central stars of PN present a WR spectrum. The spectrum of NGC 40 (Fig. 10) shows the usual narrow nebular emissions, superposed to a strange stellar spectrum, where broad emission lines are formed by strong stellar winds (see Sect. 4.4).

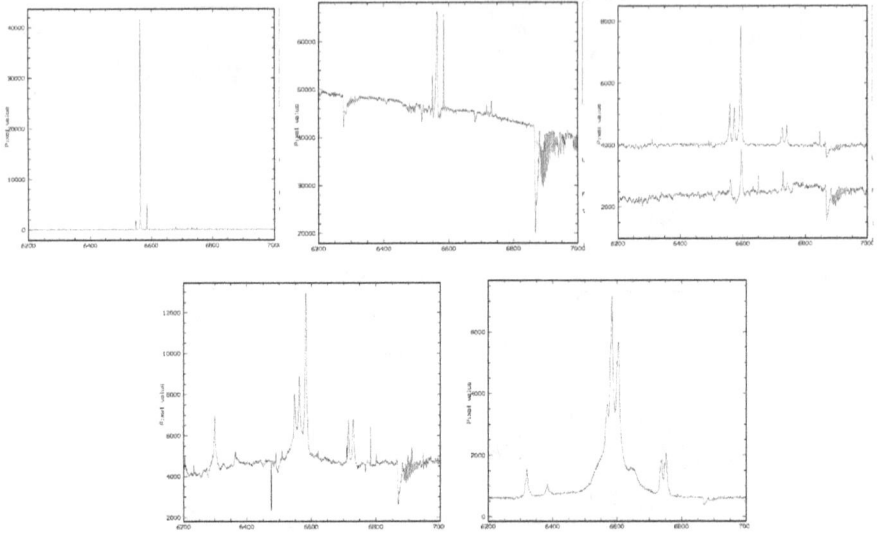

Fig. 10. (Second part). The red part of the spectrum of (4) the Orion nebula (a) exposure 30 sec – (b) a hot Trapezium star, exposure 120 sec – (4c) the spiral galaxy M 51 (upper part) and his companion (lower part), exposure 1800 sec – (5a) the galaxy M 81, exposure 1800 sec – (5b) the Seyfert galaxy NGC 4151, exposure 1800 sec. All spectra were calibrated in relative fluxes. On all spectra, note the strong absorption feature at 5870 Å due to the water molecule of the terrestrial atmosphere. The spectra were performed in March 2003 by pre-doctoral students from Strasbourg Observatory, at OHP-tel. 152 cm, AURELIE spectrograph, R = 5000.

Spectra of H II Regions. The **Orion nebula M 42** is a typical starforming H II Region. The spectrum shows the very strong Hα line, and the relatively faint [NII] and [SII] lines. The [SII] lines ratio indicate a low density of about 500 cm^{-3}.

A longer exposure of the central part of M 42 shows the stellar continuum of one of the blue stars of the Trapezium, with the Hα absorption fulfilled by the nebular Hα (saturated) emission.

Spectra of galaxies

All spiral galaxies contain H II Regions. They appear as red "blobs" along the spiral arms on color images, and are detected by narrow nebular lines on the galactic spectra. **M 51** is a possible interacting galaxies, where the small spherical companion is probably attracted by the giant spiral. The nebular emission lines show a mean redshift $z \cdot c \simeq 490 \pm 25$ km s^{-1}. The distance D of the galaxy could be estimated:

$$z \cdot c = H_0 \cdot D \implies D = 8 \pm 0.3 \text{ Mpc} \quad \text{(with } H_0 = 60 \text{ km s}^{-1}/\text{Mpc)}.$$

Note that the Hα emission line appears very faint inside the strong Hα absorption due to the dominant stellar population of the companion.

Fig. 11. Images of typical Planetary Nebulae. From left to right, downward: NGC 6543, the *CatEye nebula*, observed by the HST. IC 5148 observed by the *Neustadt Astrophotography Group*. *ETHOS 1* observed in 2009 at the VLT by Henri Boffin *et al*. The *Soap Bubble nebula* discovered independently in 2007–2009 by amateurs: D.M. Jurasevich, K.B. Quattrocchi and M. Helm, Nicolas Outters.

The galaxy **M 81** is the site of violent events. Therefore, the [SII], [NII] et [OI] lines appear very strong relatively to Hα, and the wings of all lines are broadened.

The galaxy **NGC 4151** is a Seyfert I galaxy, having an active nucleus, showing very broad lines with a high velocity dispersion.

4 Physical parameters of the nebulae. The case of Planetary Nebulae

Planetary Nebulae represent late stages of low mass stars evolution. Images obtained with small instruments often show a typical ring aspect (as IC 5148, Fig. 11).

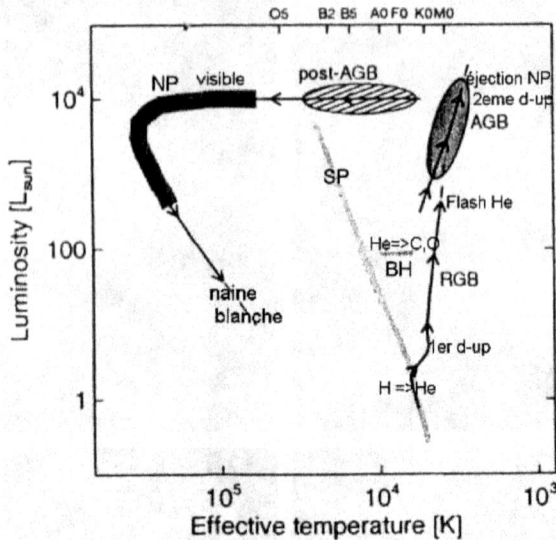

Fig. 12. Evolution of a solar mass star from the main sequence to the Red Giant Branch (RGB), the Horizontal Branch (HB), the Asymptotic Giant Branch (AGB), the post-AGB, the planetary nebula (PN), towards the white dwarf.

High spatial resolution images performed by the HST or the ESO-VLT reveal complicated structures, signatures of violent mass-loss episodes (see for example the HST images of Mz3, of NGC 6826, or of NGC 6543, and the VLT image of ETHOS 1 (Fig. 11 *left*).

Planetary nebulae (PN) are a step of the evolution of low and intermediate mass stars. They are the rapid transition objects between the *Asymptotic Giant Branch* (AGB) stars and white dwarf phases (see Fig. 12).

Let us briefly summaries the properties of an ideal PN. It consists of two interacting parts: (1) a spherical cloud or shell of gas centered around and originating in (2) the central star, which expelled the gas shell during the AGB stage. The central star is becoming a hot compressed star, radiating essentially in the ultraviolet spectral range, exciting the gas which re-radiates most of the energy through emission lines. The planetary nebula has thus a complexe morphology defined by multi-wavelenght observations (Fig. 13), and the nebular component has a well-defined spectral shape in the visible region of the spectrum (Fig. 6).

In fact, the morphology and radiation patterns can vary widely between individual objects. The morphology is affected by various processes: the ejection mechanism, the rotation of the ejecting star and its binary nature; the nebular mass and the interaction between the ejected gas shell and the interstellar medium (Franck 1994; Kwok 2001; Miszalski *et al.* 2008, 2009). Moreover, the morphology is almost unknown for a large fraction of the detected PN which have a "stellar" appearance because of large distances and/or compactness. The nebular spectrum,

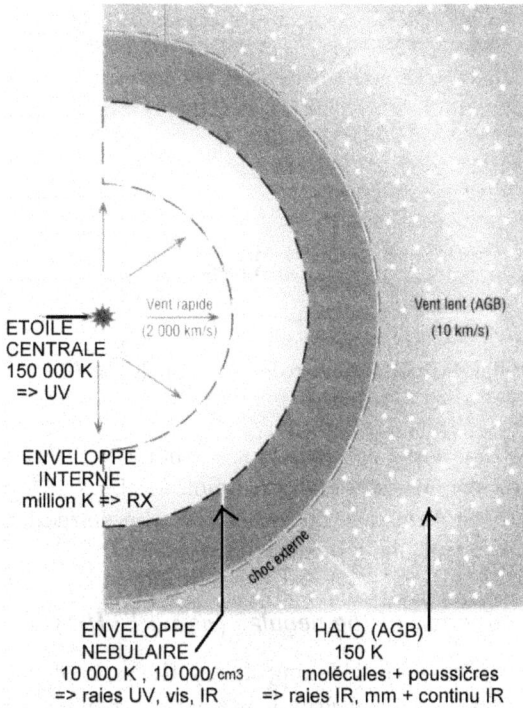

Fig. 13. Structure of a planetary nebula, observed from various points of view (http://www.lcsd.gov.hk/CE/Museum/Space/Programs/SkyShow/Butterflies/).

on the other hand, is affected by the radiation properties of the central star as well as by the physical properties of the gas shell itself and its chemical composition.

4.1 The catalogs

All observed data can be found in catalogs.

The *Strasbourg-ESO Catalogue of Galactic Planetary Nebulae* (SECGPN) was published by Acker *et al.* (1992, 1996) and its digital version is hosted by the CDS: http://vizier.u-strasbg.fr/viz-bin/VizieRPlanetary Nebulae V/84. The SECGPN catalogue was highly extended by Parker, Acker *et al.* (2006), and Miszalski *et al.* (2008), in the framework of the *MASH* project conducted by Strasbourg and Sydney universities:

(*Macquarie/AAO/Strasbourg Hα Planetary Nebulae Galactic Catalog* http://vizier.u-strasbg.fr/vizier/MASH).

A global catalog of the about 3000 PN known in 2011 will be published by the CDS, Strasbourg, see Miszalski, Acker, Parker, Ochsenbein (2011, in preparation). This catalog will include the few PN discovered by amateurs (see Fig. 11). In order to determine the properties of the observed PN, their spectra are analyzed using

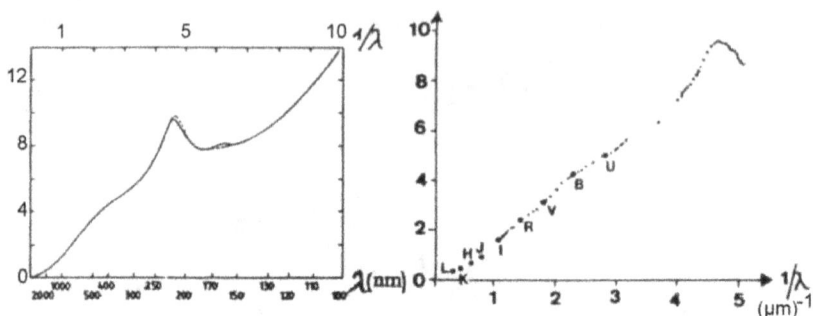

Fig. 14. Interstellar extinction and reddening coefficient (from Pottasch 1984).

the lines as plasma diagnostics, and assuming that the lines are produced in an isothermal gas at uniform density and ionization level.

First of all, the observed lines intensities must be de-reddened, and therefore it is necessary to determine the reddening for each PN.

4.2 Interstellar extinction of the nebulae (using $F(H\alpha)/F(H\beta)$)

The interstellar extinction and reddening change the ratio of the observed spectral lines; for example the relative logarithmic extinction coefficient for $\frac{F(H\alpha)}{F(H\beta)}$ is 0.325 (see Fig. 14).

The extinction constant is defined as: $c(H\beta) = log\left[\frac{F(H\beta)^{th}}{F(H\beta)^{obs}}\right]$. If we know the theoretical (th) value of a line ratio, say $\left[\frac{F(H\alpha)}{F(H\beta)}\right]^{th} = 2.85$, we can calculate c from the observed Balmer decrement:

$$c(H\beta) = log\left[\frac{(F(H\alpha)/F(H\beta))^{obs}}{(F(H\alpha)/F(H\beta))^{th}}\right] \times \frac{1}{0,325}.$$

In the *SECGPN*, the values of $I(H\alpha)^{obs}$ relative to $I(H\beta)^{obs} = 100$ are listed for each PN. We find:

$$c(H\beta) = log(\frac{I(H\alpha)^{obs}}{100 \times 2.85}) \times \frac{1}{0.325} = 3.08 log I(H\alpha)^{obs} - 7.55.$$

For a very close PN with negligible extinction we would find: $I(H\alpha)^{obs} \simeq 285$ if $I(H\beta)^{obs} = 100$. Most PN have a high extinction (more than 3 for large distances). Extinction constants were determined for a large sample of PN by Tylenda *et al.* (1992).

4.3 Distances of the nebulae

The distance determination of a PN is very difficult because none of the standard methods used in astronomy can be applied. Individual distances could be assigned

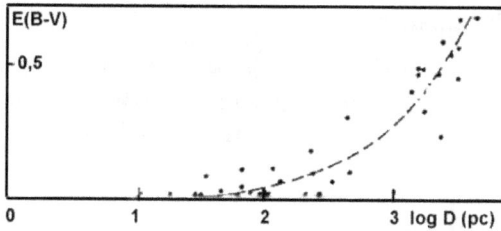

Fig. 15. Relation between the color index and the distance (pc) of stars around (15 arcmin) the planetary nebula M 1-7. The distance of the PN can be estimated from the measurement of the extinction ($c(H\beta) = 0.7$) and the derived color index $E(B - V) = c(H\beta)/1.48$.

only for a few objects: about 10 PN were observed by Hipparcos (Acker *et al.* 1998).

Individual distances (*using the extinction constant $c(H\beta)$*)
The E(B-V) color index should be determined for stars seen in a close field around the PN. For these stars, the distance can be determined thanks to parallax measurements or observed photometric data (see the Strasbourg CDS database: http://cdsweb.u-strasbg.fr/CDS.html). The distance of the PN could be estimated from the graphical relation between the color index and the distance (Fig. 15).

Statistical distances (*using the de-reddened $H\beta$ flux and the angular radius*)
- Distances can be calculated using a statistical method, the so-called *Shlovsky-method*. Assuming that all nebulae have a similar ionized mass M and are optically thin, the distance can be derived from the angular radius θ expressed in arcsec and the $H\beta$ flux corrected for the interstellar extinction, assuming a mean value of the electronic temperature of 10^4 (see Pottasch 1984, p.115):

$$D = 22.8 \times \frac{M^{2/5}}{(FF^{1/5} \times (F(H\beta)^c)^{1/5} \times \theta^{3/5})}.$$

Assuming a value of 0.75 for the filling factor FF and $M = 0.2\ M_\odot$ for the nebular mass, with $\log H(\beta)^c = \log F(\beta)^{obs} + c$, we obtain the following formula if $F(\beta)$ is expressed in 10^{-11} W.m^{-2} and D in kpc:

$$\log D = 1.11 - 0.2 \log F(\beta)^c - 0.6 \log \theta.$$

As an illustration, the distance is about 3.2 kpc if the corrected $H\beta$ flux is equal to 10^{-11} W.m^{-2} and the radius equal to $10''$.
- For compact, dense and optically thick nebulae, the mass can be expressed, either as a function of the electronic density, or as a function of the $H\beta$ flux, the distance, and the electronic density; a combination of the corresponding relations (see Pottasch 1984), gives an other distance determination, with the same units as above: $\log D = 0.48 - 0.5. \log F(\beta)^c$.

4.4   The nebular parameters

Parameters of the nebular plasma can be estimated from the de-reddened lines intensities (see Sect. 2.2). The **temperature** is estimated from the ratio of the [O III] lines intensity (see Fig. 5). The **density** is determined using the ratio of the [S II] lines intensity.

The chemical abundances

The chemical composition of planetary nebulae traces (i) the abundances of the interstellar medium from which they are formed, and (ii) the nucleosynthesis and dredge-up that have taken place during the AGB phase (Fig. 12). Therefore the study of the abundances of large sample of PN yields information both on galactic chemical evolution and on stellar evolution: the nebular abundances of oxygen, neon, argon and sulphur tell us about the metallicity (related to the age) of the star forming region at the time the star was formed, whereas carbon, nitrogen, and helium reflect the stellar nucleosynthesis depending on the progenitor's mass.

The various chemical elements present in the nebula are more or less ionized by the stellar UV photons. Each ionized element belongs to a specific *Strömgren sphere*, having a radius determined by the stellar temperature and the ionisation potential of the considered ion.

As an example, the radius of the Strömgren sphere for He^{++} can be expressed as follows:

$$\int_{4.\nu_0(H)}^{\infty} \frac{L_\nu}{h\nu} \, d\nu = \frac{4}{3}\pi R^3 N_e \left(N_{He^{++}}\right)\alpha(He^+, T).$$

The dereddened spectra are analyzed for deriving abundances, with the assumption that the nebula is represented by a homogeneous volume of constant electron temperature, density and degree of ionization. With the electron temperatures and electron densities obtained, the emissivities of all lines can be computed and thus from the observed line intensities relative to $F(H\beta)$, the ionic abundances relative to H^+ are deduced. Whenever both [N II] and [O III] electron temperatures could be determined, we use the [N II] temperature for the low ionization species (N II, O II, S II, S III) and the [O III] temperature for the higher species (O III, Ne III, Ar IV). By applying empirical ionization correction formulae, elemental abundances are derived (see Osterbrock 1974). This procedure is applied in an applet created by Dr Joachim Köppen, allowing a rapid estimation of the plasma parameters: http://astro.u-strasbg.fr/ koppen/applet.

The nebular expansion velocity

The expansion velocity is deduced from the splitting of the nebular line (having a wavelength λ) into two components separated by a value S (Å):

$$V_{exp} = \frac{1}{2} \cdot S \cdot \frac{c}{\lambda}.$$

For the Hα line: $V_{exp} = 23 \, S(\text{Å})$.

For most PN, the expansion velocity lies around $25 \, \text{km s}^{-1}$, measurable on high resolution spectra (better than $5 \, \text{km s}^{-1}$ corresponding to 0.1 Å in the red spectral range). The expansion velocity of NGC 2392 is one of the highest known

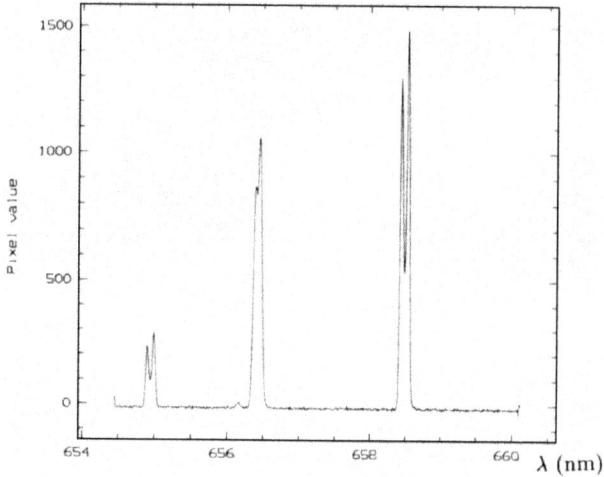

Fig. 16. High resolution spectrum of the PN NGC 6620, observed by the author at the ESO/CAT telescope with a resolution power of 60 000 (0.01 Å).

for the PN and can be measured with a low resolution as shown on Figure 9. The measurements of the separation S of the splitter lines [N II] and [S II] lead to a velocity of $85 \pm 3\,\mathrm{km\,s^{-1}}$, whereas the Hα line appears broad and not splitted. An other example is shown on Figure 16. On the red spectrum of the PN NGC 6620, the [NII] lines appear narrow and clearly splitted compared to the Hα line. The derived expansion velocities are $21\,\mathrm{km\,s^{-1}}$ for Hα and 23 for [NII].

Such a difference is explained by two considerations:

- the velocity V_{exp} is higher in the external cooler regions, where the density and the ionisation potential of the ions are low (case of the [N II] lines);

- the thermal broadening is the highest for Hα, as the thermal velocity is proportional to $\sqrt{T/m}$, with $m_H = 1$ and $m_N = 14$. Therefore the [N II] line is $\sqrt{14}$ narrower than the Hα line.

The evolutionary stage of the PN can be estimated from the *expansion age* of the nebula:

$$age(10^3 yrs) = 2.37 \times \frac{\theta \times D}{V_{exp}},$$

where the angular radius θ is expressed in arcsec, the expansion velocity V_{exp} in $\mathrm{km\,s^{-1}}$, and the distance D in kpc.

4.5 The parameters of the central star

Central stars of planetary nebulae (CSPN) are very hot ($T \geq 30\,000\,\mathrm{K}$) and their spectra are dominated by a blue continuum with only a few absorption lines. About

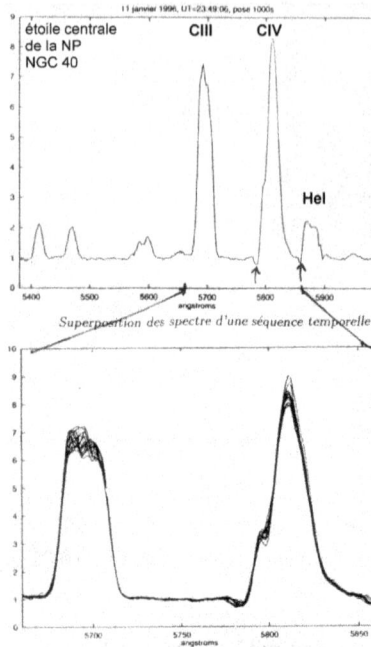

Fig. 17. The CIII-5696 and CIV-5801/5812 lines of the [WC8] central star of the planetary nebula NGC 40, observed by the author (OHP-tel. 152 cm, AURELIE spectrograph, $R = 11\,000$). The lower part of the figure shows superposed profiles obtained by a series of consecutive 15-min exposures. The top of the CIII line shows small structures with a variable velocity on relatively short time-scales, reflecting clumping of the stellar wind.

10% of the known CSPN present a [WR] spectrum (same kind of spectrum as WR massive stars, but marked [WR] for the low mass CSPN).

Stellar winds and [WR] nuclei
Strong stellar winds create around the star an expanding circumstellar shell, with an extension of some 10–100 stellar radius (some 10^6 km). The planetary nebula is a very large envelope (mean extension of some 10^{12} km) surrounding this complex stellar system (star + wind-photosphere).

For example, the CSPN of NGC 40 is relatively cool ([WC 8]-type, see Acker & Neiner 2003), and therefore presents low ionisation CII and CIII lines. The proeminent lines are the CIII-5696 and CIV-5801/5812 lines (Fig. 17). The measured FWHM of the CIII line indicate a stellar wind of about $1\,000\,\mathrm{km\,s^{-1}}$.

An absorption feature appears at the blue edge of the CIV and the HeI emission lines of NGC 40, leading to a typical *P-Cygni profile*. As shown on Figure 18, the sharp blue edge of this absorption allows to estimate the terminal velocity of the stellar wind. The large emission of the envelope (area of the H-part of the Figure) gives the value of the mass-loss.

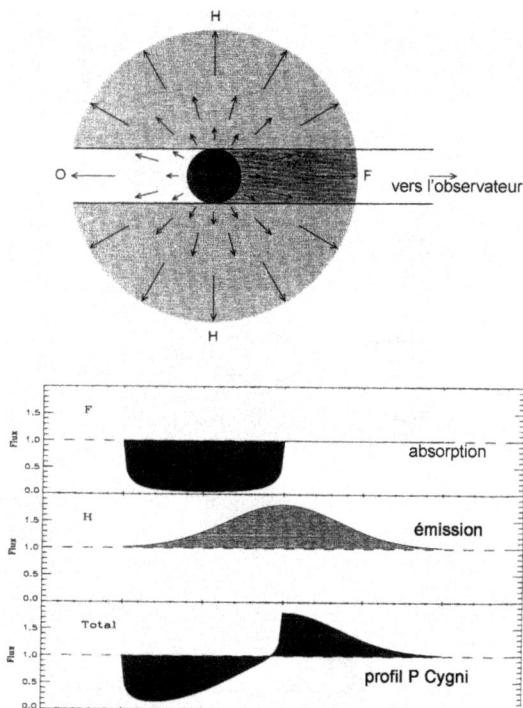

Fig. 18. Formation of a P-Cyg profile in circumstellar envelope created by a strong stellar wind (from Lamers & Casinelli 1999).

Stellar temperature T^* (*using the excitation class*)

A "cool" central star ($30\,000$ K) will be surrounded by a low ionized gas, essentially neutral; a "hot" central star (100000 K) strongly ionises the gas, which emits bright lines of [O III] and He II-4686 Å; the intensity of these lines relative to Hβ allows the determination of an *excitation class*, loosely related to the temperature of the central star T^* (Fig. 19).

Stellar luminosity L^* (*using stellar magnitude V, c, T^*, and distance D*).

- The absolute bolometric magnitude M^* of the star depends on the distance and on the stellar magnitude V corrected using the extinction constant c and the bolometric correction: $BC = -42,5 + 10\log T^* + 29\,000/T^*$:

$$M^* = V - 2,11c - BC - 5\log D + 5.$$

- If the total flux emitted by the central star is absorbed and re-emitted by the gas, the stellar luminosity can be related to the Hβ flux. An empirical relation between the stellar flux and the Hβ flux gives a median value of 160 for the ratio, covering however a wide range from about 100 to 600, or even more in the case of extremely cool central stars.

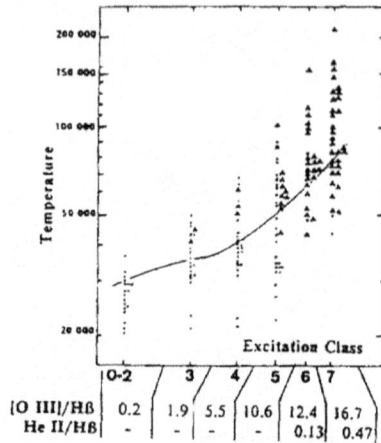

Fig. 19. Temperature of the central star as a function of the excitation class of the nebula derived from typical values of line ratios (from Acker *et al.* 1991).

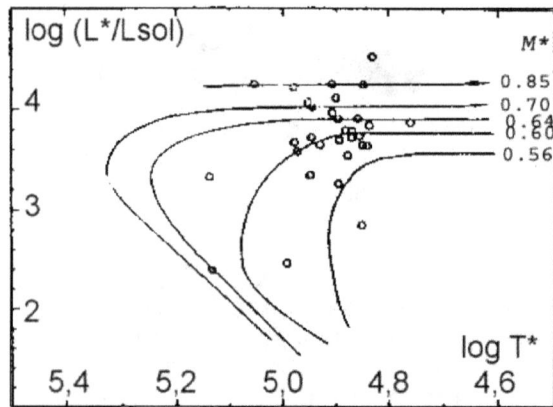

Fig. 20. Temperature/Luminosity diagram for the central stars of PN. Theoretical tracks are plotted for various values of the CSPN mass (0.56 to 0.85 solar masses), corresponding to initial masses of 0.8 to 6 solar masses for the stellar progenitor (see similar tracks in Acker *et al.* 2003; Gesicki *et al.* 2003).

Stellar mass (*from the position of the CSPN on the Temperature/Luminosity diagram*). The mass of the CSPN can be estimated by comparison with theoretical evolutionary tracks calculated for various masses (see Fig. 20). Note that the theoretical limit of the mass of the CSPN becoming a white dwarf is 1.44 solar masses.

The binarity of the central star is not easy to determine observationally, as the CSPN are very faint objects, and as the orbital period is very short in most cases. As a high proportion of stars are binaries, one can expect a fairly high rate of binary CSPN. In addition, a lot of PN show a bipolar structure, which implies the presence of a dusty disk around the star. Such a structure can be due (i) to a fast rotation of the AGB-star, leading to the ejection of an equatorial disk, or (ii) to the binarity of the central star, leading to higher density of the ejected dust in the orbital plane. About 50 binary CSPN are known in 2010, and the relation with a bipolar nebular structure is shown by Brent Miszalski *et al.* (2008b, 2009). As an example, the nucleus of the bipolar PN *ETHOS 1* (see Fig. 11) is a close binary, with a separation of 2 millions km and an orbital period of 12 hours.

5 Conclusions

Nebulae are fascinating objects: diffuse star-forming regions, circumstellar shells around protostars or degenerated stars at the end of their life, and majestic galaxies containing all these objects. Emission lines objects are the best candidates for productive work, as their emission is concentrated in specific radiations providing a lot of information.

The knowledge of the mechanisms of production of the recombination and forbidden lines lead to determine many nebular parameters:

– the interstellar extinction is obtained thanks the Balmer decrement;

– the electron temperature of the nebula is deduced from the ratio of the [O III] lines;

– the electron density is related to the ratio of the [S II] red doublet (note that the global intensity of these lines is discriminating for the separation of low density nebulae from nebulae dominated by shocks);

– the temperature of the ionizing star is related to the excitation-class of the nebula determined by the intensity of the Helium II λ4686 line.

Many thanks to Joachim Köppen for the portable spectrograph he constructed, and for his *applet* allowing quick emission lines analysis. Many thanks to C. Neiner and J.P. Rozelot for their careful regard in the final TEXpresentation. The author acknowledges the Observatoire de Haute-Provence for providing the AURELIE spectra obtained by the Strasbourg DEA/master-students in March 2003.

Références

Acker, A., 2005, Astronomie – Introduction, Masson – (Dunod Ed.) (New Edition 2012)

Acker, A., & Neiner, C., 2003, A&A, 403, 659

Acker, A., Fresneau, A., Pottasch, S.R., & Jasniewicz, G., 1998, A&A, 337, 253A

Acker, A., Tylenda, R., Ochsenbein, *et al.*, 1992, ESO Publication Strasbourg-ESO Catalogue of Galactic planetary nebulae

Acker, A., Gleizes, F., Tylenda, R., & Stenholm, B., 1991, Ann. Phys. Fr., 361

Allen, 2000, Astrophysical quantities, Fourth edition (Springer ed.)

Aller, L.H., 1956, Gaseous Nebulae (London: Chapman-Hall)

Carroll, B.W., & Ostlie, D.A., 1996, An Introduction to Modern Astrophysics, (Addison-Wesley Publishing Company)

Franck, A., 1994, AJ, 107, 261F

Gesicki, K., Acker, A., & Zijlstra, A., 2003, A&A, 400, 957

Kwok, S., 2001, Ap&SS, 265, 3

Lamers, J.G., & Cassinelli, J.P., 1999, Introduction to stellar winds (Cambridge University Press)

Miszalski, B., Parker, Q.A., Acker, A., et al., 2008, MNRAS, 384, 525

Miszalski, B., Acker, A., Moffat, A.F.J., Parker, Q.A., & Udalski, A., 2008b, A&A, 488, L79

Miszalski, B., Acker, A., Moffat, A.F.J., Parker, Q.A., & Udalski, A., 2009, A&A, 496, 813

Miszalski, B., Acker, A., Parker, Q.A., & Moffat, A.F.J., 2009, A&A, 505, 249

Osterbrock, D.E., 1974, Astrophysics of Gaseous Nebulae, Freeman & Co. (San Francisco)

Parker, Q.A., Acker, A., Frew, D.J., et al., 2006, MNRAS, 373, 79

Pottasch, S.R., 1984, Planetary Nebulae (Reidel Publish. Company, Dordrecht)

Sabbadin, F., 1984, A&AS, 58, 273

Sabbadin, F., 1986, A&AS, 64, 579

Tylenda R., Acker, A., Stenholm, B., & Köppen, J., 1992, A&AS, 95, 337

European Astronomical Society Publications Series

TITLES PUBLISHed

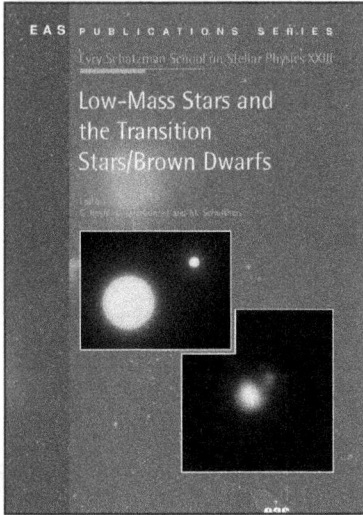

Founded in 1990, the European Astronomical Society (EAS) aims to contribute to and promote the advancement of all aspects of astronomy in Europe.

The European Astronomical Society Publications Series was launched to publish the proceedings of selected scientific meetings, including the proceedings of conferences, symposia and workshops, held in Europe and beyond.

The EAS review process ensures that only research of the highest scientific standards is accepted for publication.

Vol.	Tiles	Editors	ISBN	Price
❒ 58	ECLA - European Conference on Laboratory Astrophysics	C. Stehlé, C. Joblin and L. d'Hendecourt	978-2-7598-0941-7	69 €
❒ 57	Low-Mass Stars and the Transition Stars/Brown Dwarfs - Evry Schatzman School on Stellar Physics XXIII	C. Reylé , C. Charbonnel and M. Schultheis	978-2-7598-0819-9	44 €
❒ 56	The Role of the Disk-Halo Interaction in Galaxy Evolution: Outflow vs. Infall?	M.A. de Avillez	978-2-7598-0787-1	70 €
❒ 55	Understanding Solar Activity: Advances and Challenges	M. Faurobert, C. Fang and T. Corbard	978-2-7598-0752-9	69 €
❒ 54	Oxygen in the universe	Coordinator: G. Stasińska	978-2-7598-0710-9	70 €
❒ 53	CYGNUS 2011: Third International Conference on directional Detection of Dark Matter	F. Mayet, D. Santos	978-2-7598-0721-5	52 €
❒ 52	Condition and Impact of Star Formation – New Results with Herschel and Beyond	M. Röllig, R. Simon, V. Ossenkopf and J. Stutzki	978-2-7598-0696-6	61 €
❒ 51	Star Formation in the Local Universe	C. Charbonnel and T. Montmerle	978-2-7598-0695-9	52,75 €
❒ 50	Scientific Writing for Young Astronomers, Part 2	C. Sterken	978-2-7598-0639-3	61 €
❒ 49	Scientific Writing for Young Astronomers, Part 1	C. Sterken	978-2-7598-0506-8	36 €
❒ 48	CRAL-2010, A universe of Dwarf Galaxies	M. Koleva, Ph. Prugniel and I. Vauglin	978-2-7598-0662-1	89 €

www.eas-journal.org

Vol.	Tiles	Editors	ISBN	Price
❏ 47	Astronomical Spectrography for Amateurs	J.-P. Rozelot and C. Neiner	978-2-7598-0629-4	36 €
❏ 46	PAHs and the Universe: A Symposium to Celebrate the 25th Anniversary of the PAH Hypothesis	C. Joblin and A.G.G.M. Tielens	978-2-7598-0624-9	70 €
❏ 45	GAIA: At the Frontiers of Astrometry	C. Turon, F. Meynadier and F. Arenou	978-2-7598-0608-9	80 €
❏ 44	JENAM 2008: Grand Challenges in Computational Astrophysics	H. Wozniak and G. Hensler	978-2-7598-0606-5	27 €
❏ 43	Non-LTE Line Formation for Trace Elements in Stellar Atmospheres	R. Monier, B. Smalley, G. Wahlgren and Ph. Stee	978-2-7598-0588-4	44 €
42	Extrasolar Planets in Multi-Body Systems: Theory and Observations	K. Goździewski, A. Niedzielski and J. Schneider	Out of Print	
❏ 41	Physics and Astrophysics of Planetary Systems	T. Montmerle, D. Ehrenreich and A.-M. Lagrange	978-2-7598-0490-0	97 €
❏ 40	Non-LTE Line Formation for Trace Elements in Stellar Atmospheres	L. Spinoglio and N. Epchtein	978-2-7598-0485-6	89 €
❏ 39	Stellar Magnetism	C. Neiner and J.-P. Zahn	978-2-7598-0483-2	40 €
❏ 38	3rd ARENA Conference	M. Goupil, Z. Kolláth, N. Nardetto and P. Kervella	978-2-7598-0460-3	33 €
❏ 57	Astrophysics Detector Workshop 2008	P. Kern	978-2-7598-0441-2	64 €
❏ 36	Dark Energy and Dark Matter: Observations, Experiments and Theories	E. Pécontal, T. Buchert, Ph. Di Stefano and Y. Copin	978-2-7598-0439-9	56 €
❏ 35	Interstellar Dust from Astronomical Observations to Fundamental Studies	F. Boulanger, C. Joblin, A. Jones and S. Madden	978-2-86-883981-7	48 €
❏ 34	Astronomy in the Submillimeter and Far Infrared Domains with the Herschel Space Observatory	L. Pagani and M. Gerin	978-2-7598-0390-3	40 €
❏ 33	2nd ARENA Conference on "The Astrophysical Science Cases at Dome C"	H. Zinnecker, N. Epchtein and H. Rauer	978-2-7598-0380-4	56 €
❏ 32	Stellar Nucleosynthesis 50 years after B2FH	C. Charbonnel and J.-P. Zahn	978-2-7598-0365-1	64 €
❏ 31	Far-Infrared Workshop 2007	C. Kramer, S. Aalto and R. Simon	978-2-7598-0360-6	40 €
❏ 30	Spanish Relativity Meeting - Encuentros elativistas Españoles - ERE2007	A. Oscoz, E. Mediavilla and M. Serra-Ricart	978-2-7598-0062-9	64 €
❏ 29	Tidal Effects in Stars, Planets and Disks	M.-J. Goupil and J.-P. Zahn	978-2-7598-0088-1	48 €
❏ 28	Perspectives in Radiative Transfer and Interferometry	S. Wolf, F. Allard and Ph. Stee	978-2-7598-0074-2	26 €
❏ 27	The Third European Summer School on Experimental Nuclear Astrophysics	M. Busso, R.G. Pizzone, C. Rolfs, et al.	978-2-7598-0031-5	40 €
❏ 26	Stellar Evolution and Seismic Tools for Asteroseismology	C.W. Straka, Y. Lebreton and M.J.P.F.G. Monteiro	978-2-7598-0029-2	33 €
❏ 25	1st Arena Conference on "Large Astronomical Infrastructures at CONCORDIA, prospects and constraints for Antarctic Optical/IR Astronomy"	N. Epchtein and M. Candidi	978-2-7598-0017-9	48 €
❏ 24	CRAL-2006. Chemodynamics: From First Stars to Local Galaxies	E. Emsellem, H. Wozniak, G. Massacrier, et al.	978-2-7598-0013-1	48 €

www.eas-journal.org

Vol.	Tiles	Editors	ISBN	Price
☐ 23	Sky Polarisation at Far-Infrared to Radio Wavelengths	M.-A. Miville-Deschênes and F. Boulanger	978-2-86-883980-0	40 €
☐ 22	Astronomy with High Contrast Imaging III	M. Carbillet, A. Ferrari and C. Aime	2-86883-902-9	56 €
☐ 21	Stellar Fluid Dynamics and Numerical Simulations: From the Sun to Neutron Stars	M. Rieutord and B. Dubrulle	2-86883-928-2	56 €
☐ 20	Mass Profiles and Shapes of Cosmological Structures - IAP 2005	G.A. Mamon, F. Combes, C. Deffayet and B. Fort	2-86883-917-7	48 €
☐ 19	Stars and Nuclei: A Tribute to Manuel Forestini	T. Montmerle and C. Kahane	2-86883-913-4	37 €
☐ 18	Radiative Transfer and Applications to Very Large Telescopes - GRETA	Ph. Stee	2-86883-901-0	48 €
☐ 17	Element Stratification in Stars: 40 Years of Atomic Diffusion	G. Alecian, O. Richard and S. Vauclair	2-86883-893-6	61 €
☐ 16	Teaching and Communicating Astronomy - JENAM'04	A. Ortiz-Gil and V.J. Martínez	2-86883-881-2	45 €
☐ 15	Radio Astronomy from Karl Jansky To Microjansky - JENAM'03	L.I. Gurvits, S. Frey and S. Rawlings	2-86883-735-2	40 €
☐ 14	Dome C Astronomy and Astrophysics Meeting	M. Giard, F. Casoli and F. Paletou	2-86883-835-9	64 €
☐ 13	Evolution of Massive Stars, Mass Loss and Winds	M. Heydari-Malayeri, Ph. Stee and J.-P. Zahn	2-86883-767-0	51 €
☐ 12	Astronomy with High Contrast Imaging II	C. Aime and R. Soummer	2-86883-766-2	57 €
☐ 11	The Future Astronuclear Physics	A. Jorissen, S. Goriely, M. Rayet, et al.	2-86883-750-6	48 €
☐ 10	Galactic & Stellar Dynamics - JENAM'02	C.M. Boily, P. Patsis, S. Portegies Zwart, et al.	2-86883-701-8	37 €
☐ 9	Magnetism and Activity of the Sun and Stars	J. Arnaud and N. Meunier	2-86883-646-1	57 €
☐ 8	Astronomy with High Contrast Imaging	C. Aime and R. Soummer	2-86883-687-9	57 €
☐ 7	Final Stages of Stellar Evolution	Ch. Motch and J.-M. Hameury	2-86883-650-X	53 €
☐ 6	Observing with the VLTI	G. Perrin and F. Malbet	2-86883-652-6	43 €
☐ 5	Radiative Transfer and Hydrodynamics in Astrophysics - GRETA	Ph. Stee	2-86883-621-6	26 €
☐ 4	Infrared and Submillimeter Space Astronomy	M. Giard, J.P. Bernard, A. Klotz and I. Ristorcelli	2-86883-612-7	66 €
☐ 3	Star Formation and the Physics of Young Stars	J. Bouvier and J.-P. Zahn	2-86883-601-1	41 €
☐ 2	GAIA: A European Space Project	O. Bienaymé and C. Turon	2-86883-597-X	61 €
☐ 1	AGN in their Cosmic Environment - JENAM'99	B. Rocca-Volmerange and H. Sol	2-86883-563-5	31 €

European Astronomical Society
Publications Series

ORDER FORM

Please select and send back this order form to:

EDP Sciences - BP 112 - 91944 Les Ulis Cedex A - France
Tel.: 33 (0)1.69.18.75.75 - Fax: 33 (0)1.69.86.06.78

Library/Institution: ....

Last name: .... First name: ....

Address: ....

....

Zip-Code: .... City: .... Country: ....

Tel.: .... E-mail: ....

Do you wish to receive an invoice ? ❑ YES ❑ NO

ISBN codes	Unit Prices	Quantity	Total (€)
....	.... €	X	+ .... €
....	.... €	X	+ .... €
....	.... €	X	+ .... €
....	.... €	X	+ .... €
....	.... €	X	+ .... €
....	.... €	X	+ .... €
Orders will not be sent if no shipping rate included	**Shipping rates**		
France	+ 4.5 €		+ .... €
Rest of the World	+ 15 €		

Method of payment:

❑ By check made payable to EDP Sciences (in US$ or € only and to be attached to this order form)

❑ Please charge my credit card: ❑ Visa ❑ Eurocard ❑ American Express

Card number: ....

Card verification value: .... Expiration date: ....

(credit card accounts will be charged in (€) EUROS at the exchange rate applicable on transaction date)

Date and signature obligatory: .... Company stamp *(for institutes)*:

eas edp sciences

www.eas-journal.org

www.ingramcontent.com/pod-product-compliance
Lightning Source LLC
Chambersburg PA
CBHW061248220326
41599CB00028B/5568